D0146801

Teaching and learning about science and society

Teaching and learning about science and society

JOHN ZIMAN, FRS

Henry Overton Wills Professor of Physics
University of Bristol

Cambridge University Press

CAMBRIDGE

LONDON NEW YORK NEW ROCHELLE

MELBOURNE SYDNEY

Published by the Press Syndicate of the University of Cambridge
The Pitt Building, Trumpington Street, Cambridge CB2 1RP
32 East 57th Street, New York, NY 10022, USA
296 Beaconsfield Parade, Middle Park, Melbourne 3206, Australia

First published 1980

Text set in 10/13 pt Linotron 202 Times, printed and bound
in Great Britain at The Pitman Press, Bath

British Library Cataloguing in Publication Data

Ziman, John Michael
Teaching and learning about science and society.
1. Science – Study and teaching – Great Britain
I. Title
507 Q183.4.G7 80-40326

ISBN 0 521 23221 X

To
Bill, Mike, Joan, Pauline
Clive, Ernest, Tony, Judith,
– and all the rest of us.

CONTENTS

INTRODUCTION

It is a subject that goes by many different names, plain or fancy: *Social Studies in Science; Science of Science; Science and Society; Social Responsibility in Science; Science Theory; Science Policy Studies; Science in a Social Context; Liberal Studies in Science; Social Relations of Science and Technology; History/Philosophy/Sociology of Science/Technology/Knowledge;* etc. Let us call it, cryptically, *STS*, short for *Science, Technology, and Society*.

The diversity of names is characteristic, for it is highly diversified in content and significance. Some people would limit it to philosophical exercises within the groves of academia; others would pursue it politically into the industrial market place, the courts of justice and the councils of government. For some it is an area for dispassionate analysis; for others it is a cause for concern.

STS themes permeate the political, economic and cultural issues of our times. The whole subject is now a major factor in the equations of civilized life. Considered thus broadly, it has escaped the possibility of encapsulation in a single work.

But STS *education* – that is, organized instruction on various aspects of this general subject – is a much more specific topic. In the last decade or so, various courses of study have been tried out, with various objectives, and varying degrees of success, on a wide variety of students in secondary schools, universities, polytechnics and other institutions of higher education.* This development is, of course, itself a consequence of the general interest in STS themes in the political and cultural sphere, but manifests itself much more concretely, in the form of teaching curricula, lecture notes, textbooks and examination syllabuses. This is a topic on which it still seems possible to write in the hope of saying something useful.

* For brevity, all tertiary education will be referred to generically as taking place in *colleges*.

In many ways this educational innovation has made quite remarkable progress. But the movement for STS education is still small and weak by comparison with what it seeks to achieve. It is carried forward almost entirely by personal enthusiasm, with very little backing from the bulk of the scientific and teaching professions. It has not acquired the momentum of an intellectual tradition, nor a central core of academic experience on which its supporters can depend intuitively for guidance. There are dangerous symptoms of fragmentation, goal displacement and disharmony. It is becoming quite difficult, at times, to explain what STS education is *for*, what it should be *about*, and how it should be *taught*.

As a supporter of this movement, I have long felt the lack of an established *rationale* of STS as an educational subject. I had hoped that this might emerge from the continual hum of private argument and public debate that enlivens the community of STS teachers and scholars – but important questions concerning both ends and means seemed never to be asked clearly nor answered convincingly. Nobody seemed to be very perturbed about this; and yet I could not help feeling that our individual efforts might be concerted more effectively if we tried to state the goals to which they were directed and the principles on which they were based.

Of course, this faith in rationality of purpose in a largely incoherent social movement may be quite mistaken. It might be much wiser to leave well alone. Any attempt to construct a definite prescription for STS education by open discussion amongst its supporters would probably split the whole movement irrevocably. Superficially, there seems no consensus beyond a strong feeling that science education is in urgent need of reform in this general direction.

I began to write this book mainly to clarify my own thoughts on this very confused subject. I did not consult anyone about its form or contents, and had no real idea at the beginning how it would come out. But as I proceeded, I got the feeling that this movement has greater underlying unity than appears on the surface. The divergences of opinion are much more about practical matters – the topics to be treated, the intellectual approach, the techniques of teaching and examining – than about fundamental principles. The margin between success and failure with a particular course could more often be attributed to pedagogical factors, such as the interests of the students or the capabilities of the teachers, than to basic differences of aim. In other words, I discovered within STS education a common core of purpose, and a common fund of experience, on which the whole movement could perhaps begin to rely.

This book has thus become an attempt to bring to light the half-submerged rationale of STS education. By its very nature this cannot be defined precisely. To avoid futile disputes about minutiae, it had to be formulated very theoretically and schematically, without reference to particulars. But it is not just a survey of received opinions, blandly inviting agreement. Even though they have no greater force behind them than the weight of their own argument, I have not pulled my punches on significant issues. What we need is a dynamic educational tradition, where people concerned with science and education are stimulated to take up opposing positions on many controversial issues and yet keep in mind the deeper understanding they have in common. Again, so as not to exacerbate existing differences of opinion, I have presented the case impersonally, entirely on my own account and in my own words, without consultation or quotation of this or that authority.

This could be sufficient justification for excluding all references to published opinions or factual data. But there is also a practical reason for this very 'unscholarly' approach. The available literature is just too uneven, in relevance, scope and quality, to match the pattern and purpose of the book as a whole. Roughly speaking, the situation is as follows:

The first three chapters argue that science education, as conventionally organized in a country such as England, gives a misleading image of the capabilities and social function of science, to research scientists, to technological practitioners, to technical workers, and to the general public. This is a subject on which one might quote easily accessible statistical facts, such as the numbers of students at school and college taking this or that course and entering this or that employment, together with a smattering of wise or foolish views drawn from the very diffuse and tendentious literature on education in the natural sciences. I doubt whether there is better authority to be found there than personal experience and native wit.

Chapters 4–6 are taken up with a schematic 'model' of academic science, of the research and development system of which it is part, and of the social context in which the R & D system is embedded. The characteristic STS themes can be related to various aspects of this model, which thus unifies the subject matter of this new discipline. Here, of course, one might cite *ad nauseam* a very extensive and sophisticated literature on the historical, philosophical, sociological, political, economic, etc., aspects of science and technology; it must be assumed, however, that anyone reading a book such as this already has keys to the main gates into this intellectual arena.

In the last three chapters, however, dealing with various pedagogic approaches to STS education, as they are or might be put into practice, at various levels, in schools and colleges, for students taking various types of courses, the whole argument is much more empirical. As I have already remarked, there is practically no 'theoretical' literature on this new development, and even the simplest factual data concerning the numbers, sizes, methods and achievements of existing courses is only now becoming available. I do not think it would be fair to those people who are active in teaching about science to put any weight on this information until it has been analysed and followed up in much greater detail.

So the argument is put forward with little specific reference to its intellectual, academic or ideological antecedents or context. Nobody need suppose, however, that this book is disinterested or aloof in intention. The STS movement belongs to our own times, and to our own form of civilization. It impinges sharply on educational institutions, educationalists and teachers. I am writing out of personal experience of such teaching, personal commitment to the subject, and the friendship of a number of other people who have felt the same impact. For simplicity of exposition, I use the nomenclature and categories of the English educational system, with which I am personally familiar, but I think that these could be translated into their approximate North American or Continental European equivalents without breaking the thread of the argument. There may well be lessons also for other countries with more distant traditions, for science education has universal forms.

What is, perhaps, more significant is that the topic of this enquiry cannot be confined to one small corner of the world of education and science. It brings into question the goals of education and of science, their respective roles in society at large, the relationship between scientific and humanistic values, the balance between education, learning and research, the parts to be played by science teachers, scientists and technologists, and many other topics worthy of critical and imaginative thought.

Bristol, 3 October 1979 John Ziman

1

Science education – for whom?

1.1 The vocational imperative

Science is taught to many different people, at many different levels. The exact reasons why particular items of scientific knowledge are taught in particular ways to particular groups of students cannot always be determined, except by reference to traditional practice. For teachers and for pupils, science education serves a variety of purposes that are seldom clearly defined.

But one of the main reasons for including the natural sciences in secondary and tertiary education is that they are a necessary preparation for certain aspects of modern life. Many people need to know certain elements of science to practise their professions; many jobs cannot be satisfactorily performed without some degree of scientific knowledge. Not all science education is strictly *vocational*. Many school pupils and university students take courses in scientific subjects because they happen to be interested in them, or very good at them – that is for the same reason as they take 'useless' subjects such as history or classics. But the provision of the means for acquiring and transmitting scientific knowledge – schools, technical colleges, universities, teachers, lecturers, laboratories, research institutes and so on – would not be supported to the tune of so many thousands of millions of pounds if this were not an essential feature of contemporary civilization. The science that is needed by an advanced industrial society cannot be learnt by watching mother, sitting next to Nelly, watching '*Tomorrow's World*' or '*Horizon*' on the TV, reading the newspapers, poring over 'teach yourself' books in the evenings, or even by apprenticeship to a practical craft. Our technological civilization (for what it is worth – but that is not in question here) would slowly collapse if tens or hundreds of thousands of people were not spending some of the most formative years of their lives learning science systematically from professional teachers.

Of course, many young people have no clear idea for what career they should be diligently preparing themselves. Of course, there can be much argument over which bits of science are really essential to a particular job. Of course knowledge of some supposedly irrelevant aspect of science may prove unexpectedly valuable in later life. Of course, the 'scientific attitude' and scientific ways of thinking can be applied to advantage to all manner of practical affairs. Of course, the scientific view of the Universe and of Mankind is one of the integrating ideologies of our times. Of course, this knowledge is one of the finest and noblest achievements of our civilization. Of course, everybody needs to understand the powers and limitations of science in order to live more safely and happily with it. Of course, there is as much to learn about the precise use of language, and as much wisdom to acquire, from the study of a science as from any of the traditional humanities. And of course – this is a truism in every branch of education – if science were only taught better it would be a rewarding human experience for every pupil and for every teacher.

All these justifications and qualifications of science education are valid in themselves. It is good to be able to find so many excellent reasons for doing something so laborious as learning science – especially when it needs to be done anyway. The fundamental vocational goal of science education does not completely determine its content and style. The same end may be reached by various means, which may not be equivalent in other respects. It is important not to lose sight of these subsidiary goals in devising new educational techniques to meet new vocational challenges. In fact, this is just what this book is about.

But the vocational goal of science teaching and learning must not be played down. With the best of motives – intellectual, moral, professional, political – people with little experience of the teaching of science fasten their attention on those worthy secondary aspects. Even some teachers and lecturers, in their enthusiasm for and delight in the beauties of their subject and the insight they have gained by its study, forget how difficult it is to teach what has to be taught, and to learn what has to be learnt on the way to an actual career.

Any proposal for change in science education must be compatible with these realities. That is not to say that the way science is taught at present satisfactorily achieves its vocational purposes. On the contrary, it will be argued that many school children and college students would turn out better educated for the lives they will actually have to live if they were to be taught a little less science as such, and a little more *about* science. The practical, and entirely proper aim of preparing people for a variety of jobs where scientific knowledge is needed at various levels has been too narrowly interpreted as nothing more than the teaching of the theories, techniques, and practical capabilities of science, without reference to the context of thought and action where this knowledge is to be used.

Our starting point, therefore, is the people who need to know some science in their actual lives – mainly at work, but sometimes also at play. These needs are very diverse, not only in the various subjects required, but also in the extent to which any particular topic must be understood. The teaching of science, from the early years of the secondary school to postgraduate courses in the university, is equally diverse. Before looking for those general features to which general principles of reform might apply, let us distinguish the major groups of people for whom the modern system of science education mainly caters.

1.2 The research profession

The most exacting demand is for the training of research scientists. There are not very many of these – perhaps no more than one or two per thousand of the population. Only a few thousand of them need to be trained each year. Their work, as academics, as government research scientists, and in industry, is often very remote from immediate use. They are expensive to train, and sometimes apparently reckless in the extravagance of their research facilities. But their contribution to society, in the long run, is quite beyond reckoning.

Despite their small numbers, the production of research scientists is the dominant factor in the system of science education in every advanced industrial country. Many people regard it as a regrettable and outmoded tradition that the needs of this élite profession should dominate the education of the much larger mass of technologists, technicians and other useful people. But there is a rationale to this tradition that is more compelling than sentimental deference to high science and its mind-boggling discoveries.

The research scientist makes heavy demands on science education in several different dimensions. Scientific knowledge is cumulative. A scientist making an original investigation must have a firm base in what is already known. There is no profit in laboriously rediscovering past results. Somehow the research worker must be got to the existing frontiers of knowledge if he is to explore beyond them. His education must be *deep* – not just for 'training the mind', but to learn what needs to be known in a particular field to undertake research on a particular problem. The implications of this are discussed in §2.1.

But the totality of scientific knowledge, even in what we call a 'discipline', is far beyond the grasp of any student or any teacher. Of course, not everything a research scientist needs can be acquired by formal education. The search for relevant ideas or information in the published archives of science is itself a significant part of any scientific investigation. But even an outline of the key ideas in a mature science such as physics cannot be taught or learnt in many years of full-time study at school or college. Education for the research profession must become highly *specialized* (see §2.4) if it is to reach the necessary depth.

The frontiers of knowledge are far flung. If they are to be pushed back further in every direction, appropriately specialized research workers must be produced by the system. Science education must diversify into many special disciplines and departments, each staffed by the relevant academic experts. Science may be unified in principle by a metaphysical notion of 'validity' (§2.2), but without such a highly differentiated division of labour the advance of knowledge would soon falter.

Education for the research profession must, in its final stages, be deep, specialized, and diversified. Just how deep, in what manner specialized, and by what categories diversified can be a matter for endless debate in high academic circles. These debates need not concern us here – except that they are often resolved by a compromise that transfers the pressure to the earlier stages of education. If the science student who only knows one corner of physics is thought to be inadequately prepared for research in biophysics, then the necessary cellular biology must somehow be incorporated in his education at school. If it is thought to be absolutely necessary to spend at least four years studying quantum theory at successively more abstract levels in order to do research on elementary particles, then the required depth can be got by starting the subject in the Sixth Form. This is the machinery by which the educational needs of research scientists drive the whole system.

In addition to a body of knowledge, the research scientist needs training in the techniques of active research. By convention, this comes during preparation for the Ph.D. – essentially a professional apprenticeship which lies somewhat beyond the scope of this book. An important issue in science education is the extent to which the psychological experience of research can be simulated and anticipated at earlier stages of education, by 'discovery' methods of teaching, by the introduction of 'projects' to replace formal instruction, and so on. But this issue, also, would take us too far away from our main theme.

1.3 The technological professions

For simplicity, let us distinguish between the *scientist*, who is concerned with the *acquisition* of knowledge, and the *technologist*, whose work is the *application* of knowledge. The educational needs of the technologist for professional *practice* are not the same as the needs of the scientist for *research*. In reality, this distinction is not at all sharp. A professor of clinical medicine carries out research on a very practical subject and applies his knowledge for the benefit of his patients; an engineer designing a new bridge is simultaneously making an original contribution to human knowledge.

Modern technology, however, is pre-eminently scientific (§§5.3, 5.4). The knowledge to be applied in practice derives as much from organized research as from the systematic codification of past professional experience. In many cases it derives from fundamental research, directed simply towards the understanding of natural phenomena, without any conscious practical orientation. Advanced technology is not only scientific in spirit, relying for progress on deliberate investigations of present techniques and future developments: it is also *science-based*, drawing its theoretical rationale from the basic disciplines of the natural sciences. Thus, the aeronautical engineer designs the wing of an aircraft using the mathematical theory of aerodynamics, the plant breeder applies the principles of Mendelian genetics, and the oil prospector plans explorations on hunches derived from plate tectonics and the theory of continental drift.

The technological professions cannot do without science. All courses of technological higher education – medicine, dentistry, engineering, metallurgy, electronics, geophysics, glass technology, polymer science, fuel technology, mining, etc., etc. – include major components of basic science, either in the undergraduate curriculum or as a prerequisite to entry. Would-be doctors and dentists must study physiology and biochemistry (and also, apparently for traditional reasons, more remote sciences such as physics), mechanical engineers must know a good deal of classical physics and mathematics, fuel technologists need a thorough grasp of chemical thermodynamics and so on. These are genuine vocational needs. However little anatomy your family doctor may admit to having remembered or ever used, that is the intellectual framework which makes sense of her practical skills. Most of the day-to-day work of an engineer is covered by empirical design formulae and codes of standard practice – until faced with a problem that takes him or her back to the first principles of mechanics or mathematics.

But training for a technological career does not demand a very deep knowledge of a particular field of science. Technologists must try to understand the basic scientific principles of the techniques they apply, but their education needs to be thorough and specialized only in the practice of those techniques. The 'preclinical' sciences of anatomy, physiology and pathology are preparations for training in clinical skills, which are the real goal of medical education. The basic mathematics and physics of the engineering curriculum is subordinate to training in design. The research scientist is concerned with knowledge as such; the technologist is concerned with knowledge only as a basis for action. He or she must not tarry too long in the ivory towers of academia. A sound grasp of basic principles, some acquaintance with the current body of knowledge, and a brief introduction to recent advanced theories are all that he or she can afford to pick up on the way through to the real world.

Many more people are employed in technology than in scientific research – hundreds of thousands, rather than a few tens of thousands in a country such as Britain. This includes not only the medical and engineering professions, but a whole range of jobs in industry and government for which a science-based higher education is a necessary qualification. Indeed, the majority of graduates from the traditional scientific disciplines take up technological careers and receive their practical training on the job – managing computer systems, analysing chemical products, advising farmers, running breweries, publishing technical books, and an infinity of other professions.

Education for technological practice is much wider in scope, much more diverse in its institutional setting, than education for research. From a narrowly academic point of view, it makes more demand for quantity, but less for quality, from science education as a whole. In university faculties of science, these demands are often regarded as subsidiary to those of training for research; in technological faculties and institutions, the teaching of science is often defective for lack of contact with the research community. But the differentiation of institutions, faculties, departments and disciplines is less important than the different vocational goals, the different types of careers, for which students are consciously or unconsciously being prepared.

1.4 Technical employment

The vocational function of school science is much less definite. For many children, of course, scientific subjects are taken at school as stepping stones on the way into scientific or technological higher education. For many others they are merely elements in a general education that will ultimately be focussed on a career where such subjects are irrelevant. For any particular child, quite uncertain about his or her natural talents, inclinations, or future profession, there may be no real distinction between vocational and general education: chemistry, or biology, or physics is chosen because it is 'interesting', or 'likely to be useful', or 'what I am good at'. Indeed, the most compelling reasons may be negative: 'It's not interesting', or 'I'm no good at it' may rule out alternatives, leaving one or more sciences to be studied by default.

But an elementary undertanding of certain basic scientific principles seems almost essential for a very wide range of skilled work. The electrician and the radio repairman must know a bit about the physics of electromagnetism; the nurse and the physiotherapist must know some human anatomy; the engineering draughtsman and the computer programmer must be reasonably competent in mathematics; and the horticulturalist and the forester should not be quite ignorant of biology.

The amount of basic science that is used explicitly in such work must not be exaggerated. The excellence of a craftsman or technician lies in the skill with which he or she carries out a relatively familiar but not quite routine job, rather than in a capacity to analyse the task theoretically or to imagine an entirely novel way of doing it. This skill derives from sensitivity of hand and eye, guided by long experience, rather than from formal education or book learning. The traditional preparation for technical employment was by apprenticeship, where all that was needed of 'theory' could supposedly be learnt on the job, with very little reference to the sort of science taught in school.

For a variety of social and economic reasons this form of preparation for highly skilled technical trades has largely given way to more systematic training courses, where there may be heavier emphasis on theoretical knowledge, and hence more demand for prior qualifications in school science subjects. The traditional dividing line between a technical trade and a technological profession is now as indistinct as the boundary between research science and technological practice. Superficially, only a few additional years of academic education, leading to a higher qualification, separate the doctor from the nurse, or the production engineer from the workshop foreman. As more and more skilled trades become professionalized, and as the independent technological practitioner comes more and more under bureaucratic control in the government service or in corporate industry, these ancient class distinctions become less and less meaningful.

In any case, whether or not the demand for formal educational qualifications in science is vocationally justified, there is no doubt that a great many children take science subjects up to 'A' level* on their way into employment where knowledge of these subjects is relevant to their work. This also applies to many management and office jobs in industry, commerce and government, where it is essential to understand something of the technical and/or scientific background.

The very diversity of such employment makes it impossible to prescribe science curricula that would meet all these vocational needs in detail. What branch of physics should be emphasized at 'O' level† or 'A' level: electricity and electronics for work in telecommunications; properties of matter for the civil engineer or dental mechanic; heat, light and sound for the plumber or television cameraman? Is some understanding of the electron theory of valency an essential ingredient in the training of a chemical laboratory technician? How much biochemistry and physiology should be included in a practical course of animal husbandry for farmers? Even if such questions could be given precise answers, these could not be reconciled with one another in the very rough justice of school timetables and examination syllabuses.

Nevertheless, although science subjects are usually taught in schools and technical colleges without specific applications in mind, the fact must not be ignored that many of those who study them will be putting them to vocational use in due course. What they learn *about* science from their teachers may be just as significant for their careers as the knowledge *of* science that they acquire at this impressionable stage of their lives.

1.5 Science as general knowledge

Even in our technological civilization, everyday life depends very little

* That is, in the English school system, to a formal examination at the age of 18+.

† That is, the 'Ordinary' Level of General Certificate of Education, taken at 16+, the minimum school-leaving age in Britain.

on general knowledge. Indeed, it is quite astonishing how ignorant people can be about things in general ('What is the name of the Prime Minister?', 'In what continent is Canada?', 'When do birds lay their eggs?', etc.) without apparently impairing their capabilities in work and play. One can get along quite well, doing what the doctor tells us, without knowing the difference between viruses and bacteria, just as we can go on speaking prose without knowing the difference between nouns and verbs.

Beyond the merest mechanics of the three R's, the fundamental purpose of general education must be to fill in the background against which most people take on the daily business of life. It is obvious that science is a major component of that setting. Our civilization is as much based on the physics of energy and electricity, on the chemistry of steel and polythene, on the biology of antibiotics and contraceptives, as it is on the politics of capital and labour, the history of William the Conqueror and Oliver Cromwell, or the language of Shakespeare and Churchill. Science education in the early years of the secondary school is the main source of such basic knowledge for most people.

The sort of scientific knowledge that is 'useful' in this very broad sense would include general structural concepts, such as biological evolution, chemical bonding and physical dynamics. It should convey simple representations of the astronomical universe, the earth, solid matter, living cells and the human body. It should at the same time be linked to familiar everyday reality – weather, food, materials, machines, reproduction and illness. There can be no place in such a curriculum for technologically or academically specialized topics, except to exemplify the capabilities and limitations of particular scientific concepts or techniques.

The secondary school science curriculum not only conveys a specifically scientific image of the world and its inhabitants: it also transmits an attitude towards science and scientific expertise. The place of science in the popular culture of our times and the role of the scientist in our contemporary society are largely determined by the way in which scientific knowledge is presented in the classroom. Although most people learn very little science, and make very little direct use of what they learn, they are the silent majority whose views eventually carry much more weight than the tiny minority of research workers and advanced technologists. They too must learn something *about* science as part of their education about things in general.

1.6 Science education and science teaching

Within the educational system as a whole, science education appears as a relatively uniform and continuing process, through which each student is drawn, year by year, to successively higher levels of knowledge, conceptual grasp, technical skill, etc. This steady increase in the 'validity' of scientific education with student age is represented schematically in Fig. 1. But not all forms of employment demand the same degree of scientific competence. School children and students in higher education move out of science education at various ages, as they enter various more or less 'scientific' careers. In a natural progression according to age, we observe the majority of people leaving general science education along with all other formal schooling, at 16+. A substantial number, however, continue with school science subjects in the Sixth Form, in preparation for skilled technical employment. A smaller number, again, enter tertiary education and take scientific or technological degree courses to qualify for the higher science-based professions such as engineering and medicine. And at the most advanced and specialized levels of science education, we find a very small proportion of each generation being trained for research in the sciences they have studied.

This description of the vocational function of science education would be incomplete without reference to the training of the *teachers* of science. At every level, the intellectual demands of science are severe. No amount of pedagogical technique can hide an inadequate grasp of the principles to be taught. In terms of 'validity' this means that the science teacher should at least have passed successfully through the level of science *above* the level at which he is employed to teach. Thus, the conventional qualification for teaching to degree level is a higher degree such as the Ph.D., whilst school science to GCE A level should be taught by science graduates, and cannot safely be entrusted to teachers whose professional training has not carried them significantly beyond the A-level standard, even though this might be quite adequate for the teaching of general science in the earlier years of the secondary school.

The general system of science education must therefore make provision for training science teachers for each level. This may call for little more than studying a particular branch of science in the usual way to a more advanced standard; the higher one goes in the system, the less emphasis is laid on instruction in the arts of pedagogy. Indeed, as indicated in Fig. 1, the general rule is that the teachers at each level had passed through the next level without any special vocational differentiation before they took up teaching as a career. Thus, for example, specialist science teachers in secondary schools have usually taken university degrees in conventional science subjects, along with would-be research workers and technologists – and, unlike university lecturers, they get a year of professional training before they take up employment as teachers. The same applies even more forcibly for the academics themselves, whose postgraduate training is entirely directed towards professional research, without any reference to the undergraduate teaching for which they will in due course be employed.

This pattern of vocational training for teachers in the natural sciences derives, quite simply, from the intellectual structure of science itself. As will be shown in the next chapter, there is no escape from the hierarchical ordering of scientific concepts, where the 'validity' of knowledge at each level depends on deeper or more detailed knowledge that can only be acquired by passing through the next educational level. Quite literally, the teacher who has not passed through that further level 'does not know what he is talking about' and is therefore incompetent for his job.

But it has, nevertheless, a very significant effect on the goals of science education. At every level, the teacher's eye is fixed on entry to the level above (where he or she was trained in science) and is not always sufficiently attentive to the needs of the many students with different vocational intentions. The school science curriculum at A level, for example, comes to be thought of mainly as a qualification for admission to college, where degree courses and examinations are dominated in their turn by the goal of research. Intellectual snobbery – that is, greater esteem for abstract theory and 'validity' than for practical technique and 'relevance' – is reinforced by this characteristic feature of science teaching as a profession.

Of course, it is a grave misrepresentation of human reality to describe what happens in schools and colleges as the workings of a 'system'. There are idiosyncratic historical, political and social features of these activities that defy rational analysis in terms of purpose and function. Every country has its own peculiar pedagogic traditions, and goes about its educational business in its own very peculiar way.

But our science-based civilization is much the same all over the world. It provides – or demands – much the same range of skilled jobs in much the same proportions. Much the same scientific and technological knowledge is needed everywhere to carry out these jobs satisfactorily. The fundamental vocational purpose of science education thus imposes upon it a certain degree of uniformity that seems to match the universality of science itself. Fig. 1 is labelled to represent the English pattern of science education: change the names of the formal qualifications, distort the age scales a little, and it might be made to refer to almost any advanced country in the world.

Science education has a well-defined social function, which imposes severe constraints on its pedagogic style and institutional forms. Whatever else may be expected of it in the cultural or spiritual sphere, it must continue to turn out its cohorts of technically trained or scientifically informed people, for employment as electronic engineers and veterinary surgeons, nuclear physicists and computer programmers, foresters and chemistry teachers. This means that it must also respect the intellectual imperatives of 'valid' science, the hard core of reliable knowledge that is the ultimate justification for educating people to exercise these particular skills. In the next chapter we must consider the effects of these constraints on the spirit of science teaching as presently practised.

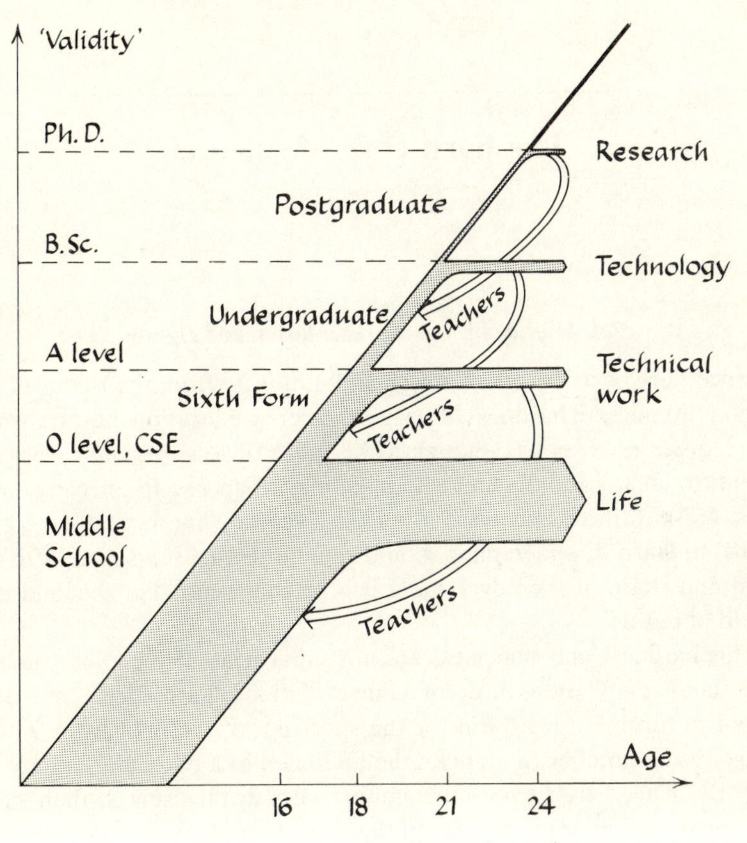

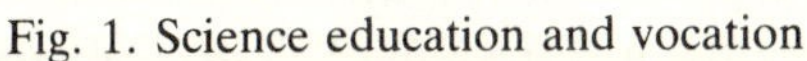
Fig. 1. Science education and vocation

2

The hard core of science

2.1 Hierarchies of representation and rigour

Science derives its practical power and authority from the rigour of its arguments and the hardness of its facts. Science education must transmit these qualities. It is not enough to be more or less acquainted with a scientific idea; to understand its meaning, or to use it correctly, one must grasp it firmly and wield it boldly. Everyone knows that science is 'hard' to learn; the metaphor should remind us that it also needs to be hard and sharp at the edge where it is to shape our thoughts and the world about us.

This hardness and sharpness are not superficial. The encyclopaedias and data compilations are, of course, full of 'hard facts', like the chemical formula for aspirin, or the spectrum of the light from Betelgeuse or the number of hairs on the abdomen of a particular variety of fruit fly. These are of no more importance, in themselves, than such historical hard facts as the date of the execution of Anne Boleyn or the Russian order of battle at Borodino. The peculiar strength of scientific knowledge is that a great many of the known facts have been organized into deeply structured patterns, from which many unknown events can be confidently inferred. History, too, has its regularities, from which much can be learnt, but there is nothing in the humanities to match the distinct categories, unavoidable necessities and reliable predictions of a well-established scientific discipline.

Metaphorically speaking, at the heart of every branch of science there is a 'map' of some aspect of the natural world. This 'map' may be very precise indeed, like the mathematical equations of mechanics and electromagnetism, or it may be a loosely defined but dominant organizing principle, like the principle of biological evolution. But somebody who has learnt a science can be expected to know about the corresponding map and to have had some practice in reading it. Scientific knowledge is as hard and sharp as the maps by which it is represented.

This does *not* mean that science has to be learnt in every detail. The important thing about a good map is that it shows only the information that is really relevant to its purpose. A good road map shows how to get from A to B, drawing particular attention to general landmarks and junctions, but excluding confusing details such as twists and turns on the way. In some circumstances a purely schematic representation, like the London Underground map, may be more useful than a 'real' map containing more irrelevant and confusing detail. Indeed, London can be quite correctly represented by a great many different maps showing railways, or bus routes, or sewers, or telephone exchanges or historic buildings, as the situation may demand.

There is no single 'scientific' map of reality – or if there were, it would be much too complicated and unwieldy to be grasped or used by anyone. But there are many different maps, of many different aspects of reality, from a variety of scientific viewpoints. The primary goal of science education, in every discipline and at every stage, is to convey the map appropriate to that stage in that discipline. Whatever may be the motives of pupils and teachers in thus coming together, this is what has to be taught and learnt.

That is not much to assert; it merely begs the question – What map is appropriate to each particular stage of each particular discipline? Should the electron theory of valency be included in O-level chemistry? How much thermodynamics do you teach to mechanical engineers? Can one hope to explain the symmetry properties of elementary particles in only 10 lectures? These are the considerations around which science curricula have to be constructed.

At every stage, the teacher can only build on what has been learnt at earlier stages. In many subjects this may not be a very severe constraint. In literary subjects, for example, the choice of a 'period', or of 'set books', for detailed study can often be left to the judgement and taste of students and teachers. The number of different topics, and the order in which they are studied, can also be decided according to arbitrary criteria.

But science education is very severely limited by the constraints of primacy and priority of topics. The different 'maps' of a subject are interconnected in a complicated pattern that can only be resolved by tracing them out in a particular order. A certain number of 'primary' or 'elementary' concepts must first be understood – for example, the classical dynamical concepts of momentum and energy, or the physiological functions of the major organs – before one can begin to understand more subtle mechanisms such as quantum-mechanical interactions or biochemical enzyme processes. These primary elements may, in fact, be drawn from the elementary maps of several different traditional disciplines, such as physics and chemistry, so that these constraints not only determine the order in which various topics can be dealt with in a particular discipline, but also tie all the scientific disciplines together at their earlier stages.

In other words, scientific knowledge has a well-defined hierarchical structure. The successive stages of science education are not only successively more subtle and difficult, matching the increasing intellectual maturity of the student growing up through the system: they also represent successive levels of generality, or of remoteness from everyday objects and events, and can only be arrived at along a path that passes through these levels step by step. The map to be mastered at each level presupposes a good understanding of the map that has been studied at the level below.

It must be clearly emphasized that this hierarchical structure is not, so to speak, totalitarian. The higher level map does not supersede the ones below it – nor even govern it in principle in every detail. Classical mechanics is good, useful, powerful scientific knowledge that is not, somehow, reduced in authority and stature by quantum mechanics. Macroscopic physiology is not rendered worthless by the marvellous discoveries of molecular biology. Reductionism – the notion that the properties of complex systems such as organisms or molecules can be 'reduced' to the laws satisfied by simpler systems such as cells or atoms – is not only a very dubious philosophy: it is a dangerous folly in science education, where the map appropriate to each level must be taught wholeheartedly according its own lights.

But many of the sharpest, hardest, and most significant achievements of science are being made at the highest levels of these hierarchies – or, as we say, in a significant inversion of the metaphor, in the *deepest* realms of theory. Not all the 'frontiers' of knowledge are, in fact, drawn on these most remote and abstract maps – but that is where a great many basic researchers must arrive and set to work if science itself is to make real progress. Every active scientific specialty penetrates in some part to these advanced levels, or acquires its own characteristic structure of successively more sophisticated conceptual schemes and techniques.

As we saw in §1.2, the education system must somehow provide a teaching structure in which at least those few people being trained for research careers can move from level to level, in various disciplines, in the correct succession, so as to arrive at the research front before they are too old and overeducated to make useful discoveries. In this part of the system, the teaching institutions, the examinations, the curricula, and the pedagogical methods must be matched reasonably well to the cognitive structure of science itself, from the barest elements to the most advanced research topics. If our educational system lacked these potentialities – that is, if some of our young people could not find ways through our schools and colleges to learn elementary particle theory, or molecular biology, or virology, or astrophysics, or general systems theory, or neural physiology, or magnetohydrodynamics, or a hundred other advanced scientific and technological subjects, then we might as well withdraw from the research game (or business) altogether.

2.2 'Valid' science

Although the number of professional scientists to be produced is quite small, the educational apparatus to train them is necessarily large and complex. The specialized sub-fields of modern science cover a very wide range, and must be expounded in great depth. The diversity of theoretical and methodological connections between the various disciplines must be opened to students moving in different directions towards active research fronts. The division of science education into disciplines and departments is mainly a convention of academic organization: the apparatus of research training, from O level to Ph.D. degree, is compactly connected, 'horizontally' by cognate and interdisciplinary subjects, and 'vertically' by the paths of students moving from level to level to the most advanced fields of research.

This educational apparatus is also unified by a common *rationale*. The theoretical concepts, the models, the experimental and observational techniques, the objects of investigation, the typical difficulties, the potential applications, and many other characteristics, differ profoundly from field to field of science. It seems absurd to associate a theoretical astrophysicist, speculating mathematically about the first five minutes of the universe, with an observational zoologist, watching chimpanzees at play. But these diverse intellectual and material activities share the same ideology, the same underlying assumptions about the possibility of discovering an order in nature, and the same criteria of credibility and cogent argument. The champions of different scientific disciplines may dispute endlessly over the fundamental significance of their various approaches towards an understanding of nature, and yet agree easily on the validity of any particular research report within its own field.

This notion of 'valid' science cannot be precisely defined. It is obviously closely related to 'hardness', or 'sharpness', or 'rigour', which we have already noted as the vital quality of scientific knowledge. But it can seldom be identified as a formal logical property of scientific discourse. It has to be assessed almost intuitively, on the basis of practical experience in research and active participation in the scientific community. In fact, to put it quite bluntly, 'valid' science is what is recognized as 'valid' by research scientists.

It follows, therefore, that the research training apparatus is not really an educational system that grows from the ground up – that is, in successive stages, each self-sufficient, matched to the slowly maturing interests, needs, and capabilities of its students. It is oriented towards, and derives its content and standards from, the final step of acquiring the deep, diverse and highly specialized knowledge needed for research itself (§1.2). It is governed by the scientific professoriate – the recognized leaders of the research community – and not, as one might have idly expected, by the leading professional science teachers at secondary and tertiary level. This authority is not confined to graduate education for the Ph.D. or to Honours degree courses in science subjects: because of the hierarchy of 'maps' to be studied at various levels, it stretches right down into the schools, to the most elementary stages of science teaching.

Let it be quite clear that this authority is not illegitimate, nor is it merely a vestige of some traditional class structure in the academic world. If people are to be trained for research careers, then they must be taught 'valid' science. This is a very long, hard road, along which one must travel as expeditiously as possible. At every stage, important, perhaps very relevant topics have to be dropped if the research frontier is to be reached in time. Within the discipline itself, it is debilitating to wallow in some comfortable pool of 'soft' knowledge acquiring vicious intellectual habits and false standards of scientific proof. It is the responsibility of the professoriate to inspect each stage on this road, and to ensure that the teaching is as 'valid' as the pupil or student can manage at that level; it is equally the responsibility of the teacher to honour that authority, and maintain the standards of 'hardness' and 'rigour' and 'sharpness' that he himself had been taught to respect.

2.3 The dominance of training for research

Yet, as we saw in chapter 1, only a small proportion of those who study science in schools, polytechnics and universities are destined for careers in scientific or technological research. How is it that their very special needs dominate the whole of science education? Why are the curricula and pedagogical methods of science teaching in all institutions, at all levels, to students taking up all manner of careers, judged almost entirely by their 'validity' as steps on the way to the frontiers of knowledge?

It is not obvious that this is a vocational necessity. On the one hand (§1.5), the value of science education for the great majority of people is in showing them the scientific 'picture' of the natural world – an altogether softer, more vivid, more evocative representation than a rigorous map. On the other hand (§1.4), the science that is really needed in many forms of technical and technological employment may be much more arbitrary and 'rule of thumb' than what would be 'valid' as a basis for more advanced training for research. For a variety of good reasons, ranging from the harsh practicalities of training engineers, doctors, and other skilled technical workers, to the high ideals of spreading scientific wisdom and understanding throughout the nation, one would expect science education to be much more diverse in goals and styles than one actually finds it.

This is fully appreciated, and often deeply deplored, by many teachers of science at every level. But the principle of 'validity' in science education cannot be dismissed lightly. There are excellent reasons that argue at length in favour of this orthodoxy, in which we are all deeply enmeshed by intellectual tradition, by institutional practice, and by rational purpose. The true spirit of science – objective, undogmatic, critical, creative – is to be found at the research front, and it is proper that science education be imbued with that spirit. Our educational system should always be open to every talent, so that it is undesirable to segregate a small number of potential researchers into special educational streams at an early age. General science that is too sloppy and technical science that is too arbitrary generate mental sets and attitudes that are damaging to individuals, to science, and to society at large. The sharpness and objectivity of 'valid' science makes it a good school subject, continually challenging in difficulty and depth, yet clearly defined in standards of performance at every level. Arguments along these lines can be elaborated at great length. They are well-founded in educational experience and in underlying philosophy. Science education is *primarily* concerned with 'valid' science, in all fields, at all levels, and for all students.

But this primacy need not be exclusive of other aims and principles of education. In particular, every young person needs to be prepared for and oriented towards the actualities of adult life and work. This preparation and orientation is as much a responsibility in science education (especially in those later stages where *only* science is being studied) as it is in the humanities and other 'value-laden' subjects.

'Validity' is the hard core principle of science. But by allowing this principle to dominate every aspect of science education, by subordinating to it the contents of every textbook, the message of every lesson and lecture, the demands of every laboratory exercise and examination question, we neglect our wider educational responsibilities. Indeed (since human nature abhors a moral vacuum) our students of science and of science-based technology are being *inappropriately* prepared, and *misleadingly* oriented towards adult actualities. The teaching of science exclusively in terms of its ultimate research status conveys philosophical, political and ethical attitudes (chapter 3) that are ill-founded, unwise, and inadequate amidst the pitfalls and gins of the real world. But before discussing these attitudes, it is necessary to look a little more closely at the content of 'valid' science education.

2.4 Theory

'Valid' science is heavily laden with theory. At each level the material to be studied is carefully ordered according to the appropriate theoretical 'map' (§2.1). In the ideal case of physics, the emphasis is on the necessity of the various 'Laws of Nature', to which all physical phenomena must, without exception, conform. Where, as in many branches of biology and geology, the subject matter of the science has many arbitrary and accidental features, formal schemes of identification and nomenclature are imposed on the mind of the student. In every scientific discipline, at every level, the aim of the teacher is to convey to his pupils an appropriate schematic representation of a wide range of experimental or observational data.

The predictive power and rationality of theoretical representations of the natural world are the glories of science. To deduce the properties of electromagnetic radiation from Maxwell's equations; to demonstrate the endless capabilities of molecular structure according to the laws of chemical valency; to explain living processes in terms of biochemical reactions; to trace out the paths of drifting continents; and to follow each branch of evolution in the geological record: these are the true delights of a scientific education.

But there is a price to be paid in suppressing diversity and contingency in order to emphasize uniformity and necessity. Our theoretical maps tend to be too idealized and simplified. The physicist's laws and equations are only strictly applicable in highly contrived and unnatural circumstances, where the elephant sliding down the plank has to be treated as a point mass and radio waves are reflected from the sea as if it were a perfectly smooth, perfectly conducting plane. Biological taxonomy abounds in disputable schemes of classification; anyway, no two biological organisms are precisely identical. Much is, indeed, known about certain physiological processes, but others are still shrouded in mystery. The great tree of life has proliferated in endless elaborate detail throughout the ages, beyond any explicit description in terms of the survival of the fittest. These deficiencies of theory are seldom made manifest in the teaching of 'valid' science until they become the subjects of active investigation for a few professional researchers.

In its avowed purpose, as a representation of reality, the schematized knowledge taught in the early stages of science education is very misleading. Those aspects of nature that fit easily into simple schemes – physical quantities such as the dead weight of an elephant – are given too much attention, whilst other, more qualitative features – like the skill with which an elephant can balance on a plank – are conveniently neglected. Indeed, as a matter of principle, all such extraneous factors must be excluded from a 'hard' science such as physics, in the name of rigour and 'validity'.

The student of a 'pure' science is being taught to see the world in terms of the theoretical map favoured by his discipline and to interpret what he observes within that single frame of reference. Since different disciplines have somewhat different theoretical frameworks, this process is often carried out in an atmosphere of scorn for other points of view. To the physicist, chemistry is thought to be 'just stamp collecting'. To the experimental physicist, mathematical theory is 'airy-fairy' – and so on. Specialization of outlook, which is a mere necessity of the division of labour in research, comes to be regarded as a virtue, rather than as a mental vice, and the administrative boundaries between the various academic disciplines are widened into intellectual gulfs.

Of course, science is seldom taught as dogmatically and unimaginatively as I have here suggested. Educational institutions are human groups, where there is much transfer of wisdom outside the classroom and lecture theatre. The tendency towards oversimplified theoretical schemes is recognized by science teachers and moderated in the learning process. Disciplinary barriers are being broken down, as research frontiers connect in interdisciplinary topics. But within the hard core of 'valid' science, as expounded in textbooks and formal lectures, there can be no concessions to accident, ignorance, untidiness, or uncertainty.

2.5 Techniques and capabilities

Training in practical techniques is a major part of science education. A student of theoretical physics or of metallurgy, of biochemistry or of botany, of ecology or of virology, of geology or of astronomy, must learn to program a computer, or prepare a microscope slide, or run an ultra-centrifuge, or dissect a leaf, or note down the behaviour of a bird, or anaesthetize a rabbit, or collect mineral specimens, or photograph a galaxy through a telescope.

Many of these practical techniques are supposed to be needed in research – which many students will never, in fact, have to do. Where these are craft skills, of no great intellectual depth in themselves, there may be some secondary educational value in the experience. An industrial manager may never need to be able to line up an optical bench or develop a photographic plate, but it will do him no harm to make this contact with recalcitrant reality. Indeed, the stubbornness with which the material world resists our clumsy efforts to subdue it scientifically goes some way to offset the naive attitudes instilled by simplified theories.

But the educational objective of laboratory 'practicals' and theoretical exercises is to illustrate and reinforce theoretical knowledge. The 'experiments' and 'problems' are carefully contrived so that they can only be done 'well' or 'badly' according as they are in conformity with the orthodox theoretical map. If the aim of the experiment is to 'verify Boyle's Law', then the student is expected to observe a series of values of the pressure and volume of the gas that can be plotted on a nice smooth curve of the correct form. If the instructions indicate that the spleen of the mouse will be found to be enlarged, then this is what the dissection should show.

There is no doubt about the educational value of such exercises in making more credible the abstract maps of theory. Laboratory work plays a very important part in generating a sense of the reality of scientific knowledge. The validity of well-established science depends very significantly on this realism. And it is only when the map at one level has been made real in this way that one can build firmly on it to reach a higher level.

The paradox of the conventional teaching laboratory in science is that although it is conceived as part of training for research, it lacks the elements of surprise and doubt that are essential to research itself. An 'experiment' whose result is already known to the experimenter is a parody of the real process. A 'problem' whose solution is to be found in a model answer at the back of the book is no more than a puzzle with closely defined rules.

Here again, there has been a movement in science education to make the student 'discover' science for himself in the practical laboratory. This movement is well intentioned, and has had very good effects in loosening up the routine details of practical classes in science. But there is no real escape in this direction from the rigours of 'valid' science; what is to be 'discovered' thus must not be different from the scientific truth which the teacher is in duty bound to transmit.

The general effect of basic training in practical scientific techniques is thus to give a false impression of the scientific enterprise. Those who enter careers where this training is not put to the test in real research or real technological development can have no personal experience of the limitations of technique, or of the fundamental importance of inventive imagination and creative criticism in the generation of new and reliable knowledge.

The same objections can be made to many of the efforts to illustrate the capabilities of science by applying it to the solution of simplified technological problems. In the early stages of engineering education, for example, the student must work through a great many exercises in which the performance of various types of machine has to be analysed according to the principles of thermodynamics. These exercises are often quite realistic in their assumptions and are perfectly valid within those assumptions. It is an important part of the work of an engineer to make such calculations. But engineering is an art as well as a science, and a full preparation for a professional career includes design projects where the student is made aware of the limitations and uncertainties of his formal theoretical techniques. Clinical medicine, also, soon escapes from the idealizations of the preclinical science subjects in the pathetic realities of the teaching hospital.

Unfortunately, not all those people who enter essentially technological employment are given a distinctly vocational education. There are many who go into careers in applied science straight after graduating from basic science courses. Others who are involved in the teaching of science to technical trainees have never had any experience either of research or of technological practice. The boundaries between education in science and education for technical or technological employment are quite indistinct, and can be crossed in either direction by students and by teachers.

The peculiar dogmatism and convergence of science education is deplored by enlightened teachers and academics. New styles of practical work, involving open-ended projects with many of the characteristics of research, are to be found in many degree courses. That is to say, the sharp transition from undergraduate to postgraduate styles of training is spread out over one or more years; the research attitude is allowed to percolate back into the Honours degree course, whilst formal instruction in the didactic mode is carried on into the period of preparation for the Ph.D. But it is very difficult to realize this educational virtue at the school level. And it may be that all we are really doing is reinforcing the 'validity' of an education directed towards training for research, as if this were the sole purpose of the system.

2.6 'Valid' science as an educational form

The long road of science education, from the early years of secondary school to the glorious achievement of ones very own Ph.D., can be an exciting and enlightening experience. For the talented student, delighting in the mastery of new skills and the discovery of yet more subtle, sophisticated and beautiful concepts, it can be exhilarating and challenging. Each new technique that we learn can contribute to the ultimate personal goal of education – to acquire a well-founded confidence in ones own powers.

As a succession of mysteries of nature is laid bare we progress to higher and higher levels of understanding. Many students stay on in science education, not because they think it will pay them to have a higher qualification, but just because they would like to learn yet more about such a fascinating subject.

For the teacher, too, there is an immense satisfaction in putting across a new concept, and watching it catch on in the minds of his or her students. The curriculum may be closely defined in its formal content, but there is always room for a new approach that will be that much clearer and more precise. It is gratifying to be in entire command of the subject, and to be able to correct every error, to answer every question, by appeal to rational argument. There is no doubt of the importance of the work: science is the keystone of our technological civilization, and those who can teach it with a proper rigour deserve well of society.

For the great mass of school children and college students, however, science education is not, perhaps, very appealing. It is a difficult subject, at which one must work quite hard – even if only to learn by heart what one cannot quite understand. On the other hand, since it is a necessary part of the training for many good careers, there is a practical incentive to passing the exams as best one can.

What I have tried to demonstrate in this chapter is that the actual *form* of science education is quite strictly determined by its content, and is not susceptible to arbitrarily large variation. In the humanities and social sciences the current educational form is largely an historical accident. Thus, for example, the question whether a degree in English should be oriented towards literary criticism (as at Cambridge) or should have a large element of linguistics (as at Oxford) seems just a matter of personal taste. But every branch of modern science eventually twigs out into fields and sub-fields of research, whose cognitive and technical style is the standard of 'validity' in that discipline. By its whole ideology and inner logic, science education is bound to establish itself as firmly as possible around this standard. The research front is, of course, in constant motion, which science education may be a little slow to follow; but there is no alternative source of legitimacy for its form and content.

It is dangerous to generalize about so diverse an educational activity as the teaching of science. Biologists and physicists seem poles apart. What is true for physical chemistry may not really apply to pharmacology. The pure basic sciences are often very reluctant to be put into the same bed with very practical technologies.

Yet there is a common ideology, a common metaphysic of high science at the research frontier which is reflected in the forms of science education at much lower levels. In all branches of science, for example, it is the hard core of 'valid' knowledge which is absolutely predominant. In every discipline, the teaching at each level is directed primarily towards 'laying foundations' for the level above (Fig. 2), with little reference to the needs of the majority of pupils who are not, in fact, going on to that higher level. All formal instruction relates to theoretical or categorical schemes derived from the hard core and validated by research, without reference to confusions, complexities or downright ignorance about the real world. Every branch of science becomes differentiated into a speciality, whose exclusiveness is held to be a sign of purity. Differences between theoretical or technical stereotypes are exaggerated for the benefit of disciplinary and professional claims of special expertise. The power of science-based techniques is emphasized; their limitations in research and in practice are scarcely mentioned.

Some of these deficiencies derive simply from human pride. Others are more specific failings of academia, and are also to be found in such humane disciplines as history, philosophy, sociology and economics. But there is one characteristic of science education, above all others, to which we must now address ourselves: it is carried on as if the historical, philosophical, sociological and economic aspects of life were quite non-existent, and unworthy of the slightest attention by a serious teacher or his dutiful pupils.

In my opinion, this is not a necessary characteristic of science education. These themes may not lie within the hard core of 'valid' science, but they are not contradicted or invalidated by the convention of teaching science as if in training for research. On the contrary, their neglect conveys to the student images of science, images of the scientist, and images of the role of science in society, which are damaging to science, to scientists, and to society itself. These misleading images will be the subject of the next chapter.

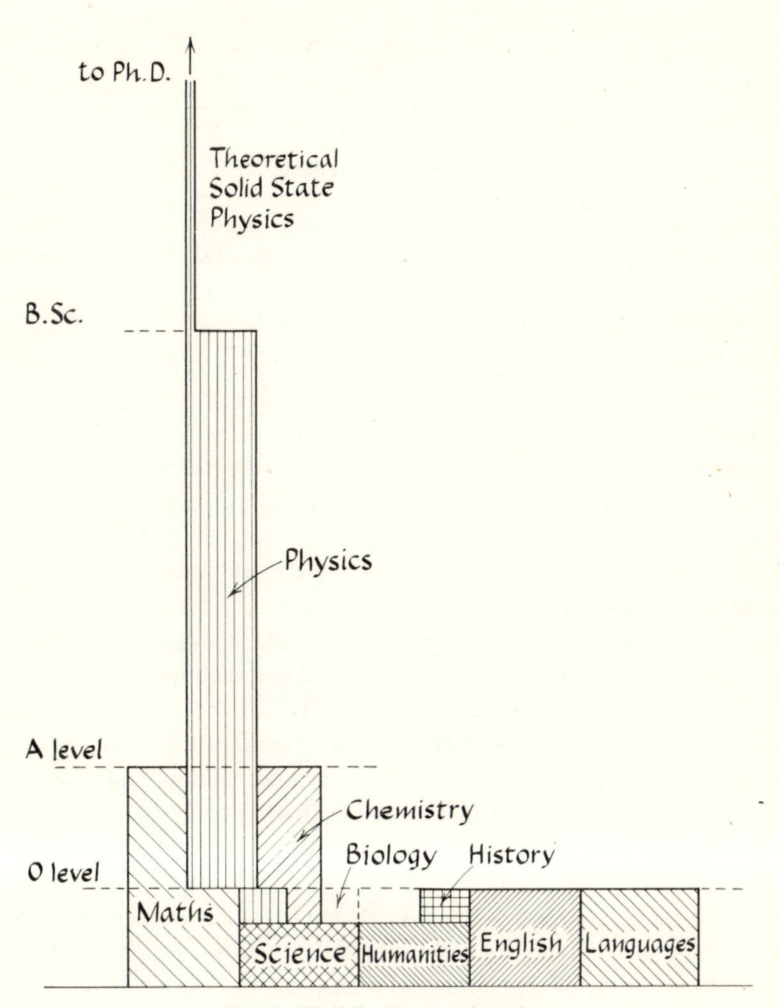

Fig. 2. 'Valid' science education

3

Scientism and its manifestations

3.1 Attitudes towards science

Scientific knowledge is so immense in extent, so extraordinarily detailed and precise, so general in its applications, that nobody can pretend that it is unimportant. One feels bound to take up a definite position towards science – even if only to be 'for' it, or 'against' it, in a simple-minded way.

This polarization of attitudes – exaggerated amongst the intelligentsia as the traditional 'Two Cultures' of humanistic and scientific education – is altogether too simple minded. On the one hand, the products of science are so very much parts of our lives that the notion of rejecting its way of thought is merely a romantic fantasy. On the other hand, those who try to let 'science' rule their lives soon find that 'cheerfulness keeps breaking in'. In practice, most people nowadays understand this pretty well when their health and comforts, or their preferences and prejudices, are at stake.

Precisely because science is so pervasive, our attitudes towards it cannot be simple and single-valued. There never could be so large and complex a human activity, so diverse in all its significations and capabilities, that had not within it a great deal of both good and evil, wisdom and folly. We can scarcely suppose that this great system of thought and action has been so perfected – or could be so perfected – that it could supersede all other sources of understanding. On the other hand, it is scarcely credible that this whole apparatus, which has transformed the world and us in it, could be a snare of fate, a cruel delusion.

The attitude to be taken towards science is an important factor in our philosophical, ethical, political, social, and economic and other opinions. But these opinions are not made much wiser by a mere antithesis of 'science' against 'anti-science'. We must attempt more subtle analyses of advantage and disadvantage, benefit and disbenefit, in many different dimensions of life. We need information about science on which to base such analyses, and occasions to exercise our arguments against informed criticism.

In other words, there is a real need, within the formal educational system, for some teaching, and learning, and discussion, and maturing of understanding, *about* science. As we saw in the previous chapter, science education traditionally repudiates any such responsibility. Its concern is solely with the content of science, as defined by the principle of 'validity'. If there is to be any such teaching, so the science educators say, then it must be under the name of some other subject, such as history, or philosophy, or ethics.

Unfortunately, the teachers of other, more humanistic subjects do not necessarily feel this responsibility, and are not always confident enough of their understanding of science to make a satisfactory job of it. It is really not quite fair for departments and faculties of science, who often have a strongly developed sense of the competence of science as a coherent method, discipline and force in human affairs, to pass this buck. It is hard to justify a separation of science teaching from teaching *about* science.

They cannot in fact be separated. By implication, in the absence of any deliberate discussion of the real situation, science education is naively 'for' science, every day, in every way, without qualification or limitation as to its reliability, scope or relevance. More by what it leaves out than by what it actually says, 'valid' science is deeply imbued with *scientism*. It reinforces, without question or comment, the widespread sentiment that science should be the only authority for belief and the only criterion for action.

Scientism is not new, and is not simply a product of our present system of science education. In various philosophical and political manifestations it is natural to our civilization, and has been a major factor in European thought since the rise of modern science in the 17th century. In its fundamental insistence on the value of scientific methods of investigation and argument, where these are applicable, it is entirely admirable. It is only really dangerous in its more extreme and doctrinaire forms, which conflict with good sense and common humanity, and are often complete perversions of science itself.

The trouble with scientism is that it takes as given an attitude 'for' science, without deeper analysis. This attitude provokes naive forms of antiscientism which are equally sterile. The very questions that are to be answered in the attempt to formulate satisfactory opinions about the role, value, use, etc., of science have already been begged.

It would be quite unreasonable, of course, to expect a teacher to preach against the very subject that he or she was also expounding. To be a good teacher of physics, say, one must have considerable confidence in the value of physics. But that does not mean that one must also be quite favourable to every application of physics, such as nuclear weapons, or that one should assert that all other branches of science are quite inferior to physics, to which, hypothetically, in the distant future, they might all be reduced. There is no necessary incompatibility between asserting the 'validity' of a particular discipline and accepting that other disciplines, other intellectual orientations and values, are also valid in their rightful places. Presumably those who teach German in English schools have deep affection for German literature and culture, but that should not blind them to some of the crimes and follies of German history, or encourage them to believe that it would be better if we all spoke German ourselves!

This chapter is concerned with the scientistic attitudes encouraged by conventional science education. But these attitudes arise more by implication than by deliberate instruction. The proper response to scientism is not antiscientism, but a deeper, more carefully balanced analysis of the ethical, philosophical or political issues that are really at stake in each particular case. Such an analysis seldom needs to be concerned with the 'validity' of the relevant scientific discipline, and in no way threatens the commitment of its teachers to that discipline.

3.2 What is known to science?

Scientific knowledge is a representation of the natural world. What is taught in a scientific discipline is some aspect of that representation – a 'map' of some local region of objects or phenomena (§2.1). The essence of 'validity' in science education (§2.4) is that this map should be very clear, precise, and connected in every part, able to give a rigorous answer to every well-formulated question within its scope.

Each discipline, with its respective map, is large enough to be relatively self-contained. The student can work in it, the scientist can research in it, for long periods without having to cross its frontiers. This region also borders on other scientific disciplines, with which science education or scientific experience has made one more or less familiar. In practice, the frontiers between the traditional disciplines are ill-defined. In some fields of chemistry, for example, one may often have to consult the books in the physics library; on the other side of chemistry the point where one enters biology is equally uncertain.

Although there is a somewhat exaggerated differentiation of research activity by academic disciplines, this does not mean that each discipline is taught as if it were the only science that existed. On the contrary, science teachers take some pride in being members of a much larger enterprise. But they are usually so specialized in their training that they have only a hazy notion of those parts of science in which they are not quite at home. The biology teacher, at every level of secondary and tertiary education, readily professes ignorance of physics, whilst the physics teacher, somewhat superior in his greater grasp of mathematics, may actually be scornful of the messy 'facts' of biology, which he feels he could never properly master.

The 'scientific world picture' is thus something that people believe in, and talk about – but it is seldom actually taught as a single topic. In the very early stages of science education some of its elementary aspects are explained, but the 'General Science' curriculum is soon divided up along traditional disciplinary lines for O level and A level. Beyond this point, science education has no place for an overall scientific representation of the natural world. Students of physics and chemistry are not taught *anything* about geology or physiology. Medical students acquire only the merest acquaintance with behavioural psychology. The academic boxes may be even smaller than a conventional discipline. In one department of physics, astronomy and cosmology may never be mentioned; in another, the physical behaviour of elementary materials such as metals and polymers may never be explained.

No doubt such matters should be left to 'general education', semi-formally in school, quite informally from newspapers, TV and personal reading. It is foolish to suppose that every student of a scientific subject could become encyclopaedic about science in all its aspects. But just because he or she is specializing in a *scientific* subject, that subject needs to be seen in its intellectual context. This is larger, and far less coherent than is normally suggested in science education.

For the existence of a single, unique, eventually-to-be-discovered scientific world picture is a myth. In and around the traditional physical and biological sciences, the local maps can be made to fit together with acceptable tolerances; beyond this region, the scientific representations shade off, and become infernally controversial. How much is known 'scientifically' in psychology? Is there such a thing as a 'social science'? Does the 'scientific world picture' have any place for man and his works? Where do human values come into this picture? These are questions that are deeply disputed. By taking for granted that there is, or could be, a unique scientific view of things, yet ignoring all questions concerning the scope of this particular view, or the legitimacy of alternative points of observation, science education is, by implication, profoundly scientistic. If science is held to be entirely valid in its own familiar regions, and if no other systems of thought are given serious consideration, the implication must be that science is the sole legitimate authority in 'mapping' the world.

It is paradoxical to suggest that by *not* trying to expound a complete scientific world picture, science education is keeping alive the myth that such a picture exists. But that, of course, is the very nature of myths – that they should not be challenged by rational investigation, and should be taught only as matters of faith. When seriously studied, they dissolve in contradiction and uncertainty. Perhaps we are all a little too sad and a little too wise nowadays to fall for *materialism* – the most naive form of scientism, which asserts that only the world picture of the traditional natural sciences is real and true, and that emotions, and ideas and social systems are mere epiphenomena of atoms and electrons and molecules and biochemistry. But the student of physics who falls in love with dogmatic psychological behaviourism, or who simply dismisses all sociology as bunk, is under the influence of that doctrine. It is the responsibility of the education system, in both its scientific and non-scientific departments, to rescue him from the ignorance that makes such folly all too easy.

3.3 How does science know it?

In a general sort of way, it is believed that science owes its power to the method by which it obtains knowledge. But science education teaches very little about this 'scientific method'. At school there is emphasis on tidiness and accuracy in writing up laboratory notes, under appropriate headings: (1) Aim of the experiment; (2) Apparatus; (3) Method; (4) Results; (5) Conclusions. The 'experiments' themselves, being performed according to a careful set of instructions ('Be careful that all rubber bungs are secured before lighting the bunsen burner') are highly contrived (§2.5), and are thus empty parodies of the true research process.

As the student approaches the research frontier, he or she may catch glimpses of the way in which new scientific knowledge is laboriously created. But until one has become an active participant in this process, one has had practically no opportunity to observe and experience it for oneself. Even as a graduate student, even as a postdoctoral research scientist, even as a lecturer or professor, one may have given no more than an idle thought, an ear to an occasional lecture, a few hours to a semi-popular book, on the fundamental question of the validity of scientific knowledge.

This is not an altogether unhealthy phenomenon. In practice scientific research proceeds very satisfactorily according to its own traditions. The research scientist picks up these traditions – criteria and standards of 'validity', appropriate research methods, formats of presentation of results and conclusions, etc., in the course of his or her apprenticeship, and learns to live within them in a way that could not be easily legitimated by philosophical argument. The actual methods of science are far more refined and subtle than any schematic 'scientific method', and its results are usually more soundly based on the good judgement of experienced research workers than could ever be justified by formal logic (see §4.5).

But university teachers who know how science really works seldom discuss this deeply with their students, and school teachers with no experience of research have only the haziest ideas on the subject. The implication of science education, in every discipline, is that what is being taught is certainly 'valid', for good and sufficient reasons which philosophers and other experts are supposed to have investigated – and therefore you had better get on and learn it properly without wasting time on such academic questions. On entry to each new level, a plausible set of arguments justifying the present state of knowledge in that field is put forward to help the student over the initial stages of unbelief, but the whole critical apparatus of doubt and difficulty is kept carefully under wraps. Of course, a clever student may not altogether repress his or her critical scepticism – but is also the first to be delighted with the possibilities of a new 'map' and to appreciate its effectiveness and 'validity'.

The notion of a general 'scientific method', which validates all scientific knowledge – and only scientific knowledge – is thus another myth that is entrenched in science education. It is conveyed by implication, as a basic metaphysical principle, and is seldom discussed or analysed. Perhaps it is realized, subconsciously, that such a direct challenge might reduce the claims of science to their proper proportions, and take away some of the mystique of science as a special source of reliable knowledge. Certainly, it would soon be discovered that the scientific methods of a discipline such as astrophysics, where instrumental measurement and mathematical speculation are closely combined, would not seem quite the same as those of cellular biology, where the truth has to be teased out of very complex situations by visual interpretations and subtle experimentation. These methods, in turn, might seem quite inapplicable to sociology or psychology. That is to say, the question whether science has, in principle, a unique 'method' would be seen to be far more subtle than most scientists realize.

Without such correctives to naive philosophical scientism, it is easy to fall into the fallacy of *positivism*, where science is regarded as the primary source of truth. In its most extreme form, 'logical positivism' even rejects all other sources of knowledge, whether of the material world or of human affairs. Thus, the power of the scientific method is presumed to extend into realms of natural and social phenomena which are, so far, quite outside the grasp of research. This extraordinary doctrine was probably more a fad of philosophers than of people with actual experience of science, and is now out of fashion. Nevertheless, science education is undoubtedly positivist in implication. Each 'valid' scientific discipline is taught as if it were perfectly true, or, if unfortunately in error, perfectly corrigible; no other manifestations of human intelligence, no other ways to truth or understanding, no other sources of belief and action, are given any serious attention.

In practice, everybody (including scientists themselves) responds to non-scientific arguments, evidences, intuitions and emotions in almost all the affairs and issues of life. The positivist programme of giving primacy to scientific forms of knowledge (whatever these are: this question itself is wide open) is relevant only to a narrow range of technological problems. But it has the effect of favouring certain kinds of information in the resolution of issues over a much wider range. As we saw in chapter 2, 'valid' science consists of a whole set of schematic representations ('maps') of the world of nature in various aspects, at various degrees of depth. A scientific problem is solved by the inspection of the appropriate map (that is, by reference to the appropriate theory, or classification scheme), where the connections and interactions are laid bare. For the positivist, every issue – the economic policy of a government, the efficacy of the penal system, the treatment of mental disorders – must be expressed as a problem situation on a suitable 'map' – an economic *theory*, a sociological *model*, a psychiatric *methodology*. Even where the existing maps are so fragmentary or speculative as to be completely misleading, he is liable to cling to them in the name of 'rationality' against common sense, practical experience, intuitive wisdom or a humane tradition. In our modern era, the advocates of almost every doctrinaire 'system', from the Jacobins of the French Revolution to the games theorists of nuclear deterrence, have justified it by reference to an idea of 'scientific method' extended far beyond its legitimate applications.

3.4 What is a science?

The public image of science is not, of course, drawn solely from formal science education. The stereotype of science for the intellectually unsophisticated person in the street is strongly refracted by the natural human taste for marvels and menaces, fed sensationally by the mass media. It would be asking too much of general science teaching in the middle school to correct this image – to provide an 'Einstein and All Them' as a companion for '1066 and All That' in the popular imagination.

But the extent to which science is misconceived by thoughtful, well-educated, socially responsible people outside the science-based professions is not to be blamed entirely on sensationalism or class ideology. 'Valid' science is peculiarly impervious to any but its own interests and criteria. C. P. Snow complained that the humanists were obstinately, scornfully neglectful of science: but the sciences are taught with corresponding scorn and neglect for all the humanistic issues that arise within their domain.

The curious student, the aspiring school teacher, or the professional scholar, with literary, aesthetic, religious or political sensibilities, is not welcomed as a temporary visitor into the citadels of advanced science. There is an academic No-Man's Land to be crossed, studded with warning notices like 'No human values please; we're scientists', 'Only science is neutral', 'Discard history before proceeding', 'Hidden minefields will explode at the first whisper of politics'. There are deep moats of fact to be swum, and great walls of theory to be scaled, mysterious passwords of terminology to be remembered, and laborious passages of experimental technique to be traversed. The inner keep of high science is as comfortable and cosy as any other billet, but it is very carefully guarded. Science education at every level is extraordinarily off-putting for the non-scientific outsider.

This isolationism of the natural sciences is not malevolent or neurotically defensive. It is a perfectly natural consequence of the doctrine of 'validity' and the tradition of teaching only the hard core of each scientific discipline (§2.2). But it means that most non-scientific intellectuals are almost as ignorant about the nature of science, its ways and means, its methods and goals, as their half-educated fellow citizens. In academia itself, where high science is supposedly most at home, there is almost as much misunderstanding and prejudice about science as in any country club or parochial church council.

Fortunately, academic institutions are somewhat loosely organized; whatever the others may think of it, each Faculty or Department is allowed great freedom in the management of its business. The sciences go their way, for good or ill, without much interference from the non-scientific disciplines. But these disciplines are themselves affected by misconceptions about the nature of science.

On the one hand, there is the antiscientific response of those who feel that science is a threat to all other branches of learning. Reacting to the primitive materialism or positivism of many scientists, they argue that scientific knowledge is evil, trivial, or false by comparison with what can be derived from other sources such as poetic insight, religious revelation, or mystical experience. They exalt a 'humanism' that is as one-sided and sectarian as the scientism against which it is an exaggerated (but often justifiable) reaction.

On the other hand, many non-scientific intellectuals, dazzled by the power and coherence of science, take it as their model for all academic activity. It is not enough for them to give the name of a science to their scholarly work: they also set themselves to do research by what they suppose to be scientific methods. They lay great emphasis on the statistical analysis of quantitative data or undertake elaborate computer simulations of the behaviour of theoretical models, as if the curiosities and contradictions of the human condition could be reduced to orderly patterns by simplified, second-hand versions of the experimental, observational and theoretical methods of elementary physics or biology. The internal validity and external evaluation of whole disciplines in the social and behavioural sciences are thus seriously warped by mistaken answers to the question 'What is a science?'.

These manifestations of scientism in academia may be expected to cure themselves, in time, by contact with reality. The ablest scholars, in the social sciences and humanities, have always been conscious of this malady. But it can be very damaging to the activities of relatively uninspired scholars and their pupils, following well-trodden paths of 'methodology' or 'formalism'. Academic scientism contributes significantly to the bewilderment and ambivalence that many people – including scientists themselves – feel about the validity of the claims of psychology, sociology, economics, and other scholarly studies of human behaviour. In the absence of a clear understanding of the status of the natural sciences as sources of 'valid' knowledge, there is no virtue in a superficial imitation of the methods of those sciences – nor, on the other hand, can grounds be established for, or against, the objective scientific point of view in the scrutiny of man as a social being. This understanding may eventually reach the scientist through his own personal experience, but can scarcely be found at all by the non-scientist within all the vast domains of 'valid' science and its conventional educational curriculum.

3.5 What can science do?

It is only a charming conceit to define science as the disinterested search for truth. The goals of research can never be completely divorced from human interest. In the highest reaches of academic science, these interests are correspondingly high-minded – the satisfaction of curiosity, the exploration of nature, the construction of an aesthetically gratifying world picture. Pure mathematics, astronomy, and elementary particle physics are quite properly justified by reference to such genuine, if unworldly, human interests.

But the motives of those who vote large sums of money for scientific research are much more mundane. They hope, eventually, to get a handsome material return on their investment. In their view, the active power of science is to be nurtured and guided towards the solution of practical problems. They accept that basic research may not, in the first instance, seem directly applicable, but are willing to support it as a strategic reserve, to be mobilized for action in due course. In other words, modern civilization rests upon the premise that science is not so much for ornament as for use.

This premise does not tell us, however, exactly how the instrument of science should be employed. It is obvious, for example, that a great variety of basic human needs have been fulfilled by science. It is also obvious that there are a great many needs still to be met. The question arises – what might science do to satisfy particular needs that we now feel to be urgent?

In its fundamentalist orientation, science education refuses to recognize the validity of this question. Attention is drawn to the entirely unpredictable and devious channels by which the discoveries of basic research are eventually put to use. How could Faraday and Maxwell have foreseen the electrical and electronic industries that have arisen from their fundamental work on the laws of electromagnetism? Is it reasonable to argue that the investigations of Mendel on the mechanism of heredity in peas should have been seen at the time to be highly relevant to the treatment of many human diseases? Given that it is impossible to calculate the relevance of any particular basic investigation to any particular human need, the whole question of how to guide science towards the solution of practical problems seems quite unanswerable.

This is a rationale of the ideology of 'pure' science, which is to be pursued by individuals motivated entirely by the disinterested search for truth, and yet must be supported lavishly by society for the benefits that will eventually float down from it, like manna from Heaven. Within limits, this is a reasonable justification for quite a lot of undirected, academic research. But how is it to be squared with the existence of technologies that are deeply involved in the basic sciences. The applications of fundamental knowledge penetrate far into academia, and research directed towards the practical realization of such applications begins to look very much like pure science in its methods and immediate objectives (see §5.3). Indeed, as we have already noted (§1.3), education in 'valid' science up to degree level is as much a preparation for a career as a science-based technologist, such as an electronic engineer or physician, as it is for scientific research.

From a technical point of view, the distinctions between pure and applied knowledge, between technologically oriented science and scientific technology, seem as arbitrary as those between the strategic and tactical aspects of military operations. In conventional science education, technology is brought under the wing of 'valid' science by looking at it solely from this 'scientific' standpoint.

This manifestation of academic scientism is almost irresistible. Each science-based technology (§5.4) – 'Engineering Science', 'Mechanical Science', 'Materials Science', 'Medical Science', etc. – is represented schematically as if it were, so to speak, a realm of nature to be explored and mapped in accordance with the appropriate theoretical principles. It is very desirable that these maps should seem consistent with the 'scientific world picture', so that heavy emphasis is laid on solved or potentially soluble problems where the fundamental principles of pure science are called upon. Thus, for example, Telecommunication Engineering is represented as the science of data transfer; the problems raised within this discipline are solved by applying the principles of electromagnetic theory and solid state physics, supplemented by further theoretical or empirical knowledge such as information theory, statistical queuing theory, circuit design principles, performance characteristics of various components, etc. In the same spirit, the infinite variety of human bodily states of health is reduced to specific categories of disease, with appropriate causative mechanisms derived from the more fundamental sciences of anatomy, physiology, pathology and microbiology, and appropriate treatments justified by reference to these and other sciences such as biochemistry.

The phenomenal successes of the science-based technologies are not to be minimized. There are no firmer foundations for professional training in these technologies than the relevant scientific disciplines. And as that training proceeds, as the student or apprentice doctor or engineer becomes more familiar with the scientific ignorance, technical uncertainties, and ethical dilemmas within which his or her profession must be practised, the scientistic model is displaced by a more realistic image. Indeed, the experienced practitioner may eventually arrive at a somewhat cynical and jaundiced opinion of the 'scientific' approach to the actual problems that have to be dealt with every day, and move towards a surprisingly antiscientific attitude.

But not all of those who pass through the system of science education go on into technological practice. Their view of those technologies of which they have little first-hand experience is strongly coloured by the attitudes of their science teachers. By referring admiringly to a few standard examples where basic research has opened the way to remarkable technological progress, and by setting contrived problems within the narrow terms of such applications, the science teacher typically presents a very one-sided view of technology. The actual capabilities of scientific technology in meeting human needs, the actual range and relative priority of such needs, the economic and political circumstances in which science is to be applied, and many other aspects of cultural and social reality are entirely ignored.

Science education is thus one of the sources of another manifestation of scientism – the belief that disease, poverty, hunger, violence and all other evils of the human condition can be done away with by the deliberate application of scientific knowledge. The most extreme form of *technological optimism* holds that everything that is technically possible (for example, the construction of artificial space colonies, or the multiple cloning of human beings) must eventually be done. This doctrine is now a little out of fashion, but it is only a modest extension of the widely held belief that anything that one would like to do must eventually prove to be technically possible. From this belief there arise many misconceived public policies, such as the 'campaign against cancer', or reliance solely upon the techniques of birth control to 'fix' the population problems of developing countries.

Here again, it is important not to fall into the contrary attitude in which all the ills of mankind are blamed upon the progress of science and technology. By failing to give a balanced response to the question 'What can science do?', conventional science education provides no rational defence against sceptical, antiscientistic pessimism. And when scientists themselves ask questions like 'Science got us into this mess; can it get us out of it?', one begins to realize the total confusion of thought that can develop in the absence of the most elementary discussion of the powers and limitations of science in relation to human values and human needs. In this intellectual and moral vacuum, even those who may be most familiar with the uncertainty and incalculability of the outcome of research turn instinctively towards an *instrumental* view of science (§6.2), as if one could always find a means to any desired end by commissioning enough research on the topic.

3.6 What is science like as a job?

Science education is dominated by the distant image of research. Yet 'valid' science says nothing about the life of the research scientist. The giants whose names are commemorated in their various 'Laws', 'Equations', 'Principles', 'Methods', 'Reagents', 'Syndromes', 'Diseases', etc., are mere stereotypes, cardboard cut-outs standing symbolically as background scenery on the stage of scientific history. To the more famous there are attached stock epithets, or dramatic incidents: Einstein was 'unworldly'; Pasteur was 'a good Frenchman'; Galileo suffered persecution; Newton sat under an apple tree. Even in institutions of higher education where the teachers are actively engaged in research, there is seldom any attempt to give undergraduates an idea of what science is like as a daily job or a lifelong career.

In the absence of reliable information (see §4.3), myths abound. Research is glamorized as an ideally free and noble life, where one can both do one's own thing in the pursuit of knowledge, and at the same time deserve well of mankind for what one has done. It is assumed that the happiest scientists are those who work in the most 'exciting' fields, and that the spiritual rewards of pure science outweigh the more material compensations of the applied scientist and technologist. Research is difficult, yes, and years of painstaking effort may not yield success, but it is carried out amongst uniquely intelligent and high-minded colleagues, deeply imbued with the 'scientific attitude'. Such a life is a worthy goal for the most talented and committed students, robustly challenging and competitive, yet gratifying in its communal spirit and high public esteem.

This is the image conveyed implicitly by the system of science education. But it is obviously so unrealistic that contrary myths abound. Research is represented as the prostitution of the intellect for the benefit of power hungry and unscrupulous corporations. The 'scientific attitude' is a sham, concealing the bitterness, the deceits, the injustices, engendered by the struggle for priority. Following the injunction to 'publish or perish', mediocre scientists produce vast quantities of useless results that completely clog up the literature. It is vain to expect to make one's name by a really original discovery, because the whole system is in the hands of an élitist Establishment clinging to outmoded concepts – and so on.

In the past, when science and technology were supposed to be entirely on the side of the angels, they probably benefited from the favourable myths; now that the gifts of science are regarded with somewhat more suspicion, a great deal of damage is probably being done by the contrary misconceptions. In either case, young people enter, or fail to enter, the research profession for reasons that are entirely erroneous, seeking or avoiding styles of life that do not really exist. Since science has always been more of a vocation than a profession, and still depends enormously on the deep personal commitment of its practitioners, this gives rise to serious problems when people discover realities that fail to match their expectations.

Perhaps there is no way of really knowing what a job is like until one has tried it. Perhaps young people are shrewder, and better informed on these matters than the formal education process could possibly achieve. In any case, this question is only significant for the very small proportion of science pupils and students who can reach the top of the academic ladder and 'go on' to do research (§1.2). Yet these, we must believe, are amongst the ablest people of their generation, whose choice of profession – as between science and the humanities, between the academic and the practical, between expertise and executive action, between industry and the public service – is a sensitive factor in the well-being of the nation. This choice is not well made in almost complete ignorance about the nature of research as a career.

3.7 What can scientists do?

What sort of people do scientists become? What capabilities do they acquire through their training and research experience? What is their place in society? These, again (§6.3), are questions that are not openly discussed in science education. They are not amenable to analysis by any of the methods of 'valid' science.

But the narrow specialization needed to reach the research frontier conveys the image of the research scientist as the ultimate in specialized expertise. To make significant progress in his own field, he must know it thoroughly, to its full depth. If, for example, he is making contributions to the theory of the electronic properties of metals, he will be supposed by his scientific colleagues to be familiar with everything that is known about this subject, experimentally or theoretically. Ignorance of relevant work by other research scientists will not be forgiven: his professional standing as a scientist is grounded on the 'validity' of his knowledge of his chosen field. Of course, this knowledge must be justifiable in a larger scholarly context – in this case, theoretical solid-state physics which is itself based on more general principles such as quantum mechanics, electromagnetism, functional analysis, and so on within the academic disciplines of physics and mathematics. But he may claim no more than textbook knowledge of all aspects of those disciplines which are not apparently relevant to his own research interest.

Strictly speaking, all that a research scientist can do (§4.6) is to speak with authority on the current state of knowledge within his field, and apply himself, as best he can, to improve that knowledge by further research. If he is of high professional standing in his subject, then he may well be one of the half dozen or so world experts at the very peak of human understanding on that particular mountain in that particular range. But scientific knowledge is divided amongst hundreds of such mountains, of which any one scientist has explored only a very few.

The scientific expert is thus the living embodiment of esoteric knowledge. His niche may seem narrow and remote within the immensity and intricacy of the scientific world, but he is perfectly at home there. However imperfectly we may understand his reasoning, we can reasonably trust his judgement on his own ground. Provided that he is kept in his place, and his social role is clearly defined, his advice on strictly scientific problems that arise in political, industrial or military affairs is surely the best we can buy.

'Valid' science thus knows only one kind of expertise – the narrow excellence of the ultimate specialist. It has no place for the skill of the technical impresario, bringing knowledge from several different fields or disciplines to bear upon an urgent issue. There is no recognition of the encyclopaedic wisdom of the academic 'gatekeeper', nor of the intuitive sensibility of the design engineer, or clinical diagnostician. The combination of sound scientific understanding and capacity for decision that is needed to lead a technical enterprise is not recognized as itself a kind of expertise. The fact that the conduct of human affairs runs largely by the exercise of judgement, based on personal experience and a strong sense of the relative priorities of conflicting systems of value, is entirely outside the scope of conventional science education.

The notion that only the scientific or technological specialist can give really reliable advice leads immediately to the ideology of *technocracy*. In this manifestation of scientism, the authority of science is given human form, in the persons of the scientists themselves. By their mastery of all that is known to science, they have become the most worthy to govern society, either indirectly as 'backroom boys' manipulating the politicians and administrators or, less modestly, by learning the various arts of 'decision making' and sitting in the seats of power themselves.

Those scientists who understand the vanity of this ideology may still insist that issues such as the future of nuclear power can be reduced to 'technical' questions, such as the energy economy of a particular fuel cycle, which can be answered 'objectively' by the appropriate experts, together with 'value judgements', such as how much risk of environmental damage is acceptable, which will be determined by political processes. Their training in 'valid' science has not shown them how deeply human values and technical capabilities interpenetrate one another, and how subtle and equivocal this can make the personal role of the scientist in all practical matters. The question 'What *can* a scientist do?' shades over imperceptibly into the question 'What *should* a scientist do?' (§6.4), for which their prolonged education in 'valid' science has in no way prepared them. These questions are not asked solely of academic scientists and scientifically trained technologists: they arise quite generally in public debate and decision at all points where science and society influence one another. In conventional science education, from general science in the lower school to the postgraduate diploma, they are scarcely mentioned.

3.8 What's the good of science?

Scientific research is infinitely laborious. Its rewards are distributed very haphazardly and are of the insubstantial stuff of fame that soon fades. Those who enter such a style of life must be buoyed up with the belief that the game itself is worth more than the candle.

At the research frontiers of high science, this belief in the pursuit of knowledge as a worthy activity in its own right is an essential psychic commitment. It may be based upon a rational assessment of the potential benefits that will arise eventually from that research, but this reference is too remote, too uncertain, to support the dedication of most basic scientists to their labours. Some will admit that they are, indeed, seeking the 'recognition' that might come with success (§4.4), but would describe their own attitude as cynical and self-seeking by comparison with the ideal of the 'honest seeker after truth'. Without such a transcendental ideal, science would quickly be corrupted by their blatant self-interestedness.

The tendency to value scientific knowledge for its own sake thus arises as a practical necessity of day-to-day scientific life. For the days, or weeks, or months, or years that scientists must devote themselves, body and soul, mind and heart, to the study of the behaviour of this obscure species of ant, or the synthesis of this unimportant chemical compound, or the solution of this highly contrived mathematical problem, they must be supported by faith that what they are doing is all for the good. To suggest otherwise is to strike deep at their morale. All their fervour has been displaced and projected upon this particular task whose present outcome must seem of little significance but whose successful accomplishment must be beneficial in the long run.

This point of view is naturally generalized to include the whole of a discipline, or the whole of science itself. It spreads to students and science teachers, far beyond the research laboratory. This is the supreme manifestation of scientism, assigning to the activity of research a *moral* value that exceeds the intellectual force or the practical utility of its products. Science becomes something more than a *philosophy* or an *ideology*: it takes upon itself the trappings of a *religion*, beyond the reach of rational justification or criticism.

The most extreme manifestation of scientistic religiosity is to project the worthiness of science upon those who practise it. Research scientists are supposed to acquire (or be born with) peculiar virtues of saintliness and wisdom called 'the scientific attitude', which especially befits them for leadership in the affairs of this wicked and stupid world. This nauseating doctrine, going beyond the utilitarian argument for technocracy, was quite fashionable in the 1930's – until, as Robert Oppenheimer put it, the physicists had 'known sin' by making an atom bomb. It was never publicly repudiated by the scientific community, but has been sufficiently discredited by external events.

In practice, scientists seldom follow this religion blindly. They must find their way through conflicts of morality, in science as in the other affairs of life. Very few are so obsessed with research that they put it above the demands of making a living, bringing up a family, serving their country or taking an occasional holiday. As their material, domestic, and social responsibilities weigh more and more heavily upon them, they often find that they can give less and less of themselves to the pursuit of knowledge in all its purity – and are stricken in conscience at their fall from inward grace.

But this is not the impression conveyed by science education. Scientific knowledge is presented, without comment, as material to be valued without separate justification. Those who have produced this knowledge are to be praised for this action alone, without reference to motives or circumstances. The question, 'What's the good of science?', is not asked; but if it were, the teacher's answer would have to be: 'Science *is* a good in itself, regardless of why, and how, it is done'.

Many of the issues associated with the phrase 'social responsibility in science' (§6.4) arise from this excessive commitment to science as a good in itself. Thus, for example, unethical practices in human and animal experimentation are often claimed to be justified by the 'search for the truth', without reference to the potential utility of the knowledge to be gained. In the same way, research topics may be pursued beyond the limits of safety, courtesy, or human decency, just to satisfy the curiosity of the scientists involved. It is even possible for scientists to hire themselves out for research on the most inhumane of weapons, such as napalm, whilst asserting their moral innocence in cant phrases such as 'It's all good chemistry, after all!'. Some of the issues that are put under this heading, such as the misuse of advanced technology, or the unforeseen consequences of scientific progress, involve many other ethical, moral, political and ideological considerations (§6.5). But the tendency of science education to assign a high *absolute* value to science, which is otherwise presented as 'value free', or 'morally neutral', is the root cause of many present complaints about the irresponsibility and inhumanity of science and of scientists.

3.9 What do most scientists really believe?

Those who live by practising and teaching science can rightly protest that it is quite unjust to blame science education for all these misconceptions about science, and all the crimes and follies that these generate. They do not themselves hold these extreme scientistic attitudes, and certainly do not teach them. In fact they take very good care not to take any particular philosophical, ideological, political or religious point of view in the classroom or research laboratory. 'Valid' science does not necessarily promote the more absurd manifestations of scientism, even if it does nothing to combat them. Science education strives continually to preserve a position of neutrality in conflicts over human needs, capabilities, and goals.

Nevertheless, one might reasonably ask an experienced research scientist or teacher what he or she really believes about such matters. What, for example, is said to pupils, students, or assistants, informally, out of school, in response to the questions raised in this chapter? Where does the balance of wise opinion about science now lie? It might be somewhere along the following lines:

(a) *Science is not an end in itself.* It is only to be valued as a means towards the satisfaction of various human needs, both material and spiritual. The search for truth is not an absolutely privileged activity: there are occasions when it must give way to other moral imperatives, such as respect for life, for beauty, for justice, or for charity.

(b) *Science is not the only source of knowledge* that is relevant to our lives on earth. The wisdom of the poet, the prophet, the artist, the lawyer, the humanistic scholar, or the statesman, may give us better guidance, or deeper insight than any formal scientific analysis.

(c) *Scientific knowledge is never absolutely objective* and cannot be known for certain to be true. It is not generated by mechanical robots, but by human beings, with all their failings of blindness and prejudice. It cannot be validated by rigorous logic, only by the exercise of personal judgement. It is always to some extent subject to the material interests, the historical experience, the cultural traditions, the social relationships of those who create it. Even where it seems soundly based and uncontroversial, accepted scientific knowledge may still contain significant errors.

(d) *Scientific knowledge is reliable only over certain aspects of the natural world;* its strength lies principally in those aspects that are studied in the physical and biological sciences. And even in these traditional disciplines many valid questions remain unanswered, or are tacitly ignored, because they do not seem to come within the grasp of the 'art of the soluble'.

(e) Even within the traditional natural sciences, *it is far more difficult to arrive at a reliable scientific answer than is imagined by most people*, who know only about the great triumphs of research and discovery. For every successful investigation there are dozens that fail to reach a convincing conclusion. The main principles of physics, chemistry and biology may well be satisfactorily understood; but the work of elucidating the details of all the phenomena that are supposed to be governed by these principles is lengthy, laborious and grotesquely incomplete. Much of our science-based technology works well enough, not because it was designed rationally from first principles with a real understanding of what was going on, but simply because it has been tested in practice by old-fashioned methods of trial and error.

(f) *In many spheres of rational knowledge*, dealing with many observable aspects of the natural world – especially the individual and social behaviour of biological organisms – *there is an almost complete lack of reliable, fundamental theory*. Many particular facts are known, but there is inadequate evidence to support and make fully convincing a general set of rigorous principles analogous to the 'laws' of the physical sciences. Every claim to have discovered such overriding principles (from which strictly verifiable predictions would follow) must be regarded with very cautious scepticism.

(g) Notwithstanding many successes in the technological applications of the physical sciences and applied mathematics, the *behaviour of any complex, strongly interacting system cannot be accurately predicted over a long period*. Quantitative data processing, an understanding of underlying processes, and the capacity for strict logical deduction, are all helpful in the control of any complex system such as a space-craft, a chemical plant or a national budget. But such control cannot be maintained without continual monitoring, and unforeseen crises can only be handled by the light of experience and good sense.

(h) Every form of social action is constrained by imperfect knowledge of the situation, the short time available for cogitation, and the multitudinous possibilities of wickedness and folly. *Success in political decision-making depends on a diversity of skills and insight*. Talents such as practical experience, moral rectitude, empathy, honesty, patience, idealism, cunning, charismatic authority, etc., usually prove far more effective (for good as well as ill!) than the best knowledge available to science.

(i) The career of the research scientist seldom includes situations where rapid decisions must be taken under conditions of uncertainty, moral ambivalence, or the conflict of irreconcilable interests. For this reason *scientists are amongst those persons in society whose experience least prepares them for the most demanding responsibilities of politics, business or war.* Although some scientists do, in fact, show these talents in practice, there is no case for forcing this social role onto all scientists simply because they are engaged in producing socially powerful knowledge.

This brief assessment of the limitations of science in its social context is questionable at every point. On every possible issue that it raises, there is a wide range of serious opinion. It is, in itself, not a 'neutral', 'value-free', 'objective' assessment, for it strongly asserts a liberal, pluralistic, political and social ideology. But that happens to be the ideology of the social system within which this book is written; this would be the natural centre around which the opinions of *our* scientists might be expected to range. If it were deemed to be the responsibility of the system of science education in an industrialized democratic country to inculcate a precise doctrine concerning the philosophical and social significance of science, then this might be a reasonable outline of such a doctrine.

This, of course, is not our purpose. In an open society there is no place within the educational system for such a programme of indoctrination. The point is simply to show how very far the real opinions of thoughtful scientists and science teachers are from the scientistic views that have been sketched out earlier in this chapter. On almost every substantial issue, they would probably find themselves taking up a moderate position, neither unreservedly 'for' nor implacably 'against' science. Scientism is a more serious disorder amongst those who have been taught only a little science than it is for those who teach it.

But the challenge to conventional science education is whether it is really neutral concerning the philosophy or ideology of science. By its austere concentration on objectivity and 'validity', by its disciplinary specialization and fragmentation, by the problems that are chosen for investigation, by its examples of technological application, by its glorification of the intellectual achievements of its creators without reference to their human qualities, and in many other ways, science education conveys an attitude towards science that is very different from the moderate, eclectic, pluralistic attitude that has been sketched above. It warns, by implication, against 'soft', subjective, value-laden, humanistic points of view, without revealing the converse weaknesses of 'hard', objective, value-free, scientific analysis. Even if it is not overtly scientistic, it effectively insulates the student from any serious criticism of the naive materialism, primitive positivism and complacent technocracy that seems to be the official public face of science.

This is why there is an urgent need to teach more *about* science in schools and colleges. It is not a question of radically subverting established scientists and technological institutions, nor of diverting the main stream of science teaching from its traditional channels. The health of the scientific enterprise depends upon people having a much more accurate picture of science and technology than they get from the existing curriculum. They need to look inside the Black Box that encloses science, conceived as an instrument of social action (Fig. 3). This should be the fundamental objective of the movement for STS education – not to replace conventional science education nor to modify it out of all recognition, but to correct its unconscious bias with complementary themes.

But how should these themes be orchestrated and performed? In what key can these diverse tunes be made to harmonize, and by what instruments should they be given voice? In the next three chapters we shall consider the topics that arise when we begin to think seriously about science in its social context, and show how these can be organized into coherent structures and significant forms.

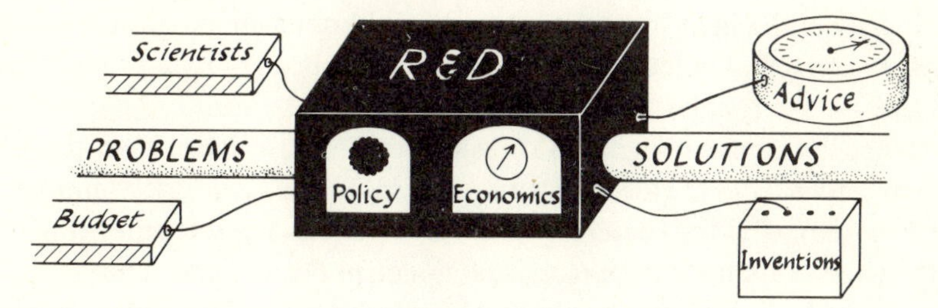

Fig. 3. The instrumental model of science

4

The social model of academic science

4.1 Unifying STS education

'Valid' science is taught as if it were unconnected with the world about it. In reality, it is linked in many ways to society, especially through its technological applications. The basic need in science education is to teach about Science, Technology, and Society, and the various ways in which they interact with one another. At this stage in the discussion let us not argue about whether this is a convenient name for a heterogeneous collection of subjects for teaching and research, or whether it should be regarded as a distinct academic discipline (see chapter 9). All that we can say is that there exists a significant movement for STS education whose main objective is to reform, or improve, or complement conventional science teaching in this general direction.

But what, in fact, does this subject consist of? For many of the advocates of STS education it seems sufficient to point out all the fallacies in the scientific attitude, and to correct them by reference to alternative points of view. Thus, for example, scientific positivism is to be opposed by philosophical arguments such as those of Karl Popper, or the notion of the objectivity of the scientific expert is to be examined critically in the light of historical evidence about technological decision-making. Each manifestation of scientism is to be combatted separately, on its own particular front.

Since the movement for STS education is little more than a loose alliance covering a wide spectrum of opinion, this piecemeal approach to the subject is conveniently non-controversial. Each teacher can feel free to draw up his own personal catalogue of horrors and errors, and to expound his own brand of alternative wisdom. The only unifying principle is the cheerful assumption that everybody is fighting in the same good cause against the scientific 'Establishment' and its out-of-date doctrines.

But this alliance quickly falls apart under adverse circumstances. Conventional scientists and science teachers still regard the whole movement with considerable suspicion. Superficially, it seems united by a negative *anti*scientific prejudice: all that the enthusiasts for STS education seem to be doing is attacking the concept of 'valid' science in the name of a variety of other causes which are not themselves well-founded and which are often inconsistent with one another. The diverse ethical, philosophical and political arguments which are called upon to justify STS education do not seem to combine into a coherent ethical, philosophical and political point of view, and are then easily defeated in detail. Put on the defensive, the STS movement shows signs of sectarianism and fragmentation. The extreme views of one or another group seem to taint the whole enterprise and provoke counter-claims from another wing. This disharmony infects the relationships between different professional groups involved in STS education – between academics in universities and polytechnics, or between school teachers and educationalists – who happen to find themselves emphasizing different aspects of the subject to students preparing for quite different careers.

Anyone who is already firmly attached to some all-embracing dogma need feel no cause for alarm in this familiar situation. For the Marxist or anti-Marxist zealot, for example, STS education is simply another battleground on which the clash of ideologies is to be fought out. Short-term success or failure may depend on the strength of tactical alliances against particular enemies, but in the end the victory of one's own doctrine must be the primary goal.

But most of the supporters of STS education lack such reach-me-down convictions. Like most thoughtful people in modern democratic societies, they are broad-minded in their ethics, eclectic in their philosophy, and pluralistic in their politics. They are disheartened by the apparent incoherence of the movement, both in practice and in principle, and yet they know that they must not surrender themselves, or their pupils, to any extreme or over-simplified doctrine merely for the sake of intellectual consistency.

What is missing is a realistic *positive* picture of science, to replace the scientistic stereotype against which they are reacting. So much of STS education seems to be concerned with what science is *not*, what it does *not* know, what it *cannot* do, or what scientists must *not* do; so little is said about what science *is*, what it *knows*, or what it rightly *does*. Science is defined, so to speak, by its negative aspects, by the social mould in which it has been cast, as if there were nothing more inside than a machine for producing valid scientific knowledge – a machine characterized more by its various weaknesses than by its actual capabilities.

The undeniable fact is that by almost any standard this machine is extraordinarily productive and efficacious. The instrumental power of science (§6.2) is not an illusion, and is not at all easily harnessed to 'socially responsible' action (§6.1). Scientistic doctrines (chapter 3) have great appeal, and technocracy is a major political force. To achieve credibility, STS education must acknowledge these positive features of science, and try to show how they really arise.

This is essential if the STS theme is ever to be fully incorporated in the system of science education. There is fertile ground for the seeds of reform. Those who live by specialized research and orthodox teaching have become a little uneasy about what they are doing. They know that philosophy of science is moving away from the complacencies of positivism. They are well aware of attacks on the morality or integrity of science, and cannot deny that the stereotype of 'valid' science is not quite consistent with their personal experience in the research laboratory, or field of battle, or classroom. They too are unconsciously seeking a new image that will help them understand their life and work, even if it is not quite so comforting as the simple faith that all science is true and all scientists must be good people.

4.2 What *is* science, anyway?

For its zealous supporters, for its vague sympathizers, and even for many of its opponents, the most important theme in STS education is raised by the question 'What *is* science, anyway?'. This question cannot be answered uncontroversially. It arouses some of the deepest issues of philosophy, sociology, psychology and politics. It is not just a matter of definition, or the recognition of a distant category of being, like a scientific question such as 'What is an electron?', or 'What is a gene?', or even 'What is an elephant?'. One could learn a great deal about a person's political views or philosophy of life by their response to such an open question.

Nevertheless, in the last few decades some progress has been made towards a better characterization of science than is implied within conventional science education. This answer is not so crisp and apparently precise as a confident reference to the 'scientific attitude' (§3.8) and 'scientific method' (§3.3); but scientists themselves will recognize that it is much closer to the science they know. The older picture was so idealized and abstract that it seemed quite divorced from reality; the contemporary model of science is much more natural, and can account for many features of science as actually experienced today.

The strange thing is that this model is much more generally accepted and taken for granted than would appear from the STS literature. At first sight this literature seems hopelessly contradictory and chaotic. In its more academic reaches, STS education draws in philosophers, sociologists, social psychologists, political scientists and a host of other disciplinary specialists, all apparently tackling entirely different problems with entirely different techniques. Most people concerned with STS studies are acquainted with the theories of Robert Merton, Thomas Kuhn, Karl Popper, Michael Polanyi, and others, who seem eternally at loggerheads. They seldom appreciate that this apparent dissension conceals a fundamental agreement that science must be represented by a *social* model. In the field of 'science studies' there is now a striking convergence of underlying assumptions about what science is and the way it achieves its remarkable successes. Detailed inspection soon reveals that many supposedly conflicting opinions are not really incompatible, since they are applied to quite different aspects of science as a whole.

To the earnest disciple of one or other of these master scholars, this must seem altogether too optimistic. But I am not just pleading for liberal tolerance and eclecticism ('let a hundred flowers bloom'), just to damp down internal conflicts within the STS movement. It is, in fact, quite easy to describe the social model of science that is implicit in most of the STS literature, and to show how the standard modern theories can be incorporated within it. Although the details of this model are still a matter for research and argument, the basic outlines seem perfectly clear, and are not all that complicated or abstruse. Indeed, it is so natural and simple that those who use it accept it tacitly, as if all that must be just too obvious. It is only as one listens to debates about the validity of this or that partial opinion ('Kuhn v. Popper', etc.) that one realizes how much need there is for a clear exposition of these simple, obvious facts.

STS education must surely strive to describe science as it really is. In the general hubbub of ideologies and disciplines this is not easily made out. It is all too easy (and much more exciting!) to emphasize conflicting points of view, and to neglect those matters on which there is general agreement. But what is obvious and elementary to the scholar or teacher is not necessarily obvious to the inexperienced and immature student. The value of the social model of science is that it is a coherent and more or less accurate representation of just such matters. It is certainly far from the whole truth about science, but it can be accepted by almost every STS teacher as a general framework for the discussion of more sophisticated and controversial issues.

In any case, even if there is no agreement on the general form of the answer to the question 'What *is* science?', the social model is useful as an ordering scheme for the major topics and issues that actually arise in teaching *about* science. In its simplest form, as set out in the present chapter, it is mainly concerned with basic research as a more or less self-contained activity oriented towards the production of knowledge. In other words, it is applicable primarily to *academic* science. The scientist is seen in relation to the natural world by observation and experiment, in relation to ideas in the realm of scientific facts and theories, and in relation to other people in the scientific community. Even in academia the production of scientific knowledge is not simply a matter of experimenting and theorizing by individuals: it is a social process, in which critical interactions between scientists, and communal acts of validation play a vital part. The consensual knowledge that accumulates in the public scientific archive provides, in its turn, the intellectual framework for further research by individual scientists.

This model is very helpful in explaining both the strengths and the limitations of 'valid' scientific knowledge. But of course it greatly over-emphasizes the so-called *internal* factors in the research process. As every schoolboy knows, modern science is not a self-contained, self-sustaining activity. Even as a description of very pure and basic research, the model sketched out in this chapter must be considered only a very inadequate approximation.

But because it concentrates on the way that scientists do their work, and relate to one another, the restricted model of academic science can easily be extended to cover the whole *research and development system*. Thus, in chapter 5, we shall be able to schematize the effects of *external* social forces on scientists and on their institutions. The familiar questions that arise concerning the place of science (that is, of science *and* technology) in society are discussed in chapter 6, in the light of this extended model.

Of course, this separation into factors internal to and external to the 'scientific community' is somewhat arbitrary. In reality, all the components of the 'R & D' system are highly interdependent, and interact with one another so strongly that they cannot be dissected out and studied in isolation. Alternative schemes of analysis, laying the stress on different elements and linkages, can certainly be proposed and made plausible. But by actually making a systematic tour through a general model of this kind we encounter the whole range of topics that ought to find a place in STS education.

In my opinion, none of the multifarious aspects of the subject touched upon in these three chapters should be entirely neglected; all the key words italicized in what follows ought to appear somewhere in the total STS curriculum.

4.3 The academic scientist

Scientific *knowledge* is the product of scientific *research*, which is an activity carried out by *scientists*. For much of their work, modern scientists collaborate in research *teams* (§5.2), and make use of expensive apparatus and technical services provided by *research institutions*, such as universities and government laboratories. But in the academic approximation to the social model, the elementary components are supposed to be simply individual scientists. In principle, such a person is responsible for his or her own research work, from choosing a problem for investigation right through to getting recognition for successful results; like many other aspects of the model, this is only a zero-order approximation to a much more complex reality.

On a day-to-day basis, research calls for the solution of a never-ending sequence of technical problems – experimental, instrumental, mathematical or conceptual. The professional scientist is educated in a scientific or technological discipline, and trained to *solve research problems* in a special field. To a considerable degree, success as a research worker depends upon the skilful exercise of technique, in the same manner as any other professional worker such as a physician or computer programmer.

On a longer term, however, month by month, or year by year, the goal of research is not just to solve prescribed problems by well-defined technical methods. The job of the academic scientist is to make an *original contribution* to knowledge. Research work is a series of investigations of natural phenomena, to discover, or measure, or observe, or explain such features as may be thought significant or explicable. Psychologically it calls upon such personal traits as curiosity, obsession with ordered thought, and conceptual imagination. Research is not, of course, merely detective work, following up such factual clues as are left lying around: it usually involves direct interference with natural phenomena, by deliberate *experiment*, with its theoretical counterpart of rational *prediction*. Science is active rather than contemplative, and as much concerned with intellectual synthesis as with analysis.

Although the tactics of research in a particular field are often quite orthodox, and may be learnt as a formal technique, strategic decisions depend on experience, theoretical insight, and other forms of what Michael Polanyi has called *tacit knowledge*. Thus each scientist must exercise personal judgement in the choice of research topics, in the formulation of the questions that are to be answered, or in the methods to be used. In practice, as we shall see, these decisions are very much narrowed by the existing state of knowledge (§4.6), by the relationship of the scientist with his or her colleagues (§4.4), and by the technical resources (§5.2) available. But these constraints cannot be embodied in an exact formula or rule of thumb. According to this model, academic science depends upon there being sufficient freedom for at least some scientists to carry out research along lines that they thus deem promising, without detailed reference to any other authority than their own commitment to the pursuit of knowledge.

The goal of research is not, however, to satisfy the scientist's personal curiosity or taste for problem-solving. The results of each investigation must be communicated to the scientific community. It is the professional duty of every scientist to prepare research reports or *primary communications* for *publication* in books and learned journals. In this way, the research work of individual scientists is brought together as a body of *public* knowledge, for the benefit of science and of society at large. Since the professional career of a scientist depends upon the quality and/or quantity of these communications, he has both a responsibility for their accuracy and an incentive to make them appear as persuasive and significant as possible. For this reason, a scientific paper seldom chronicles the actual process of research, but presents a rationalized record of observations and arguments that would seem to lead firmly to a particular conclusion.

Even in his or her primary role as a research worker, the scientist has to carry out many different tasks, combining technical ability, disciplined imagination, creative energy, and skill in communication. In practice, there is a place within science for people of very diverse talents and interests, excelling in one or another of the many traits that are needed. Although the conventional qualification for a professional research career is successful performance as a student of a specialized branch of valid science, this gives little weight to some of these valuable traits. Indeed, as we shall see, this model of science makes allowance for the fact that scientists, however well chosen and trained, are of very unequal abilities in research.

4.4 The scientific community

Although primary research is carried out by independent scientists, or teams of scientists, scientific knowledge is much more than the sum of their separate discoveries. Valid science is peculiarly reliable because it is the product of a *scientific community*. This does not necessarily take the form of an organized institution such as a learned society; in principle an *invisible college* is open to any person claiming to be making an original contribution to knowledge in a particular field; in practice membership is confined to scientists with appropriate academic qualifications and research experience. From the point of view of any individual scientist, the scientific community consists simply of all the other scientists who might be interested in the results of his or her research.

It is the responsibility of the scientific community, through its learned societies or by the initiative of individual members, to create and maintain the *communication system* of science. This consists of learned journals for the publication of primary research reports, conferences where scientists can meet in person to discuss current progress in their subject, and a variety of abstract services, review journals, scholarly monographs, etc., where the research results claimed by various individual scientists are indexed and collated.

The communication system of science is not merely a medium for the public diffusion of scientific knowledge: it also sets standards for the acceptability of primary research claims. Each research report submitted for publication in a reputable learned journal is scrutinized by the editor or by *referees* who are scientifically qualified to assess its originality and validity. Since the only persons with such qualifications are other scientists in the same field, the published research of every scientist is subject to a process of *peer review* and must therefore reach a certain level of technical performance, logical argument, and coherent exposition.

But this is only the first stage of a continuing process of communal evaluation and validation of research results. It is the public duty of every member of the scientific community to study carefully the work of other scientists, and to draw attention to any errors of fact or fallacious arguments. This process of *criticism* is encouraged by *competition* between scientists in the same field; there is credit to be gained by making significant improvements on the research work of others, reinvestigating doubtful results, establishing the facts more precisely, or extending the research into new fields. In any case, current research claims are subject to careful re-assessment in review articles, progress reports, and other critical communications, often by the leading *authorities* on the subject. By means of such organized scepticism, scientific knowledge is tested and its validity accredited by the scientific community.

By convention, every published research paper must contain some element of novelty: it must not repeat what is already well-known. Although several different scientists may be near to making the same discovery, it is attributed to the first one to report it publicly. Competition for *priority* in scientific discovery thus maintains standards of *originality* in research. The scientist who makes an important new observation or theory is rewarded in various ways – by recognition of his standing as a leading scientist, by the award of a prize, or by promotion in his job. In a general way, science functions by exchanging *recognition* for *communication*.

The scientific community is not organized in a formal bureaucracy. In principle, all scientists are co-equals, with equal access to the media of communication and criticism. In practice, there is a steeply graded hierarchy of esteem, in which the leading authorities – that is, those to whom the most important discoveries are attributed – form a highly influential élite. It is they, for example, who govern the formal institutions of the scientific world, such as learned societies, and who also head the many informal *invisible colleges* covering the many different branches of scientific knowledge. Nevertheless, the ideal of a universal scientific community, embracing scientists of all countries, of all political persuasions, and of a wide range of age and experience, is an essential feature of the social model of science in its academic form.

Such a community could not function if its members did not respect certain rules of behaviour – for example, accepting the criticism of referees, or not turning to personal abuse in a scientific controversy. The competitiveness of science has to be curbed and harnessed by a set of *norms*, such as those suggested by Robert Merton. These norms are not formally prescribed, but are usually picked up by scientists in the course of training for research.

4.5 Scientific knowledge

Scientific knowledge is to be found in the scientific archives – that is, in learned journals, books, maps, computer tapes, etc., where the results of research are reported. Although a great deal of wisdom can be gleaned by private communication with individual scientific workers, the knowledge that science regards as *valid* must be available for public scrutiny. In principle, academic science has no place for esoteric, or secret information, except as it may arise incidentally during the research process.

But the information in the scientific archives is of very uneven quality. Observations or theories that have just been reported for the first time in primary papers may eventually turn out quite fallacious. It is only after successive stages of *criticism*, such as the repetition of experiments testing against alternative hypotheses, or by logical comparison with other relevant research results, that they may be regarded as well-established. Even the most cherished scientific ideas, incorporated in standard textbooks and taught as if unquestionable, have been shown to be wrong. Scientific knowledge can never claim to be absolutely true: it is always, in principle, corrigible.

Nevertheless, well-established scientific knowledge is very reliable. This is not because all scientists use an infallible research *method*. The information on which science is based is of immense diversity – instrumental measurements, visual inspection, photographs, mathematical calculations, formal reasoning, hypothetical models, inferences from previous research, experimental tests, and so on – and of very uneven logical rigour. It is impossible to construct completely watertight arguments, proving the necessary truth of a scientific idea, from such material.

The validity of a scientific fact or theory is based upon the extent to which agreement can be achieved between scientists. In practice, it is the *consensus* of the scientific community that guarantees its reliability. This consensus is not forced, but must be obtained in the face of organized scepticism according to high standards of critical argument. Science has therefore developed a number of intellectual criteria and devices by means of which consensus may be achieved reasonably promptly. For example, it is concerned almost entirely with those aspects of the natural world that can be established objectively – that is, by the agreement of many independent observers or experimenters.

The problem of arriving at a scientific consensus is particularly acute for *theories* – that is to say, conceptual schemes in which many facts are ordered into significant patterns. Experience has shown that much the most convincing demonstration of the validity of a theory is that it should make a successful *prediction* of a previously unknown fact. This is not always easy to achieve in practice, but a good demarcation principle for the initial acceptability of a scientific theory is Karl Popper's criterion that it should be potentially *falsifiable* by a subsequent experiment or observation. Thus, although scientific theories are often of great subtlety and generality, they can say practically nothing about the validity of judgements of value, as in most political and aesthetic issues, where a free consensus of opinion is seldom available.

Although scientific knowledge is not based upon any specific metaphysical assumptions, it does, in fact, rely upon various general principles that are almost universally shared, such as the existence of a coherent and intelligible external world, and the primacy of elementary logical reasoning. Thus, science can be regarded as an extension into all manner of highly specialized and particular situations of our everyday experience of the natural world.

But science is not a complete representation of the natural world, and is always in a dynamical process of reformulation and improvement of its ideas. This continual revision affects mainly the more speculative theories at the research frontier, but there are occasions when the whole structure of accepted theory in a particular field of science undergoes a conceptual revolution, or where an unexpected discovery opens up an entirely new realm of research. It is noteworthy, nevertheless, that although the *interpretation* of established scientific facts may then change remarkably, the facts themselves, and the empirical relations between them, may remain as reliable as before.

4.6 Back to the scientist again

The individual scientist is under severe competition to produce contributions that are both original and valid. He or she is also well aware of the large amount of scientific knowledge that already exists in almost every field. In these circumstances there is little option but to try to add just a little bit to the contents of the scientific archives. One accepts the orthodox *paradigm* of the subject, and undertakes what Thomas Kuhn has called *normal* science. This is apparent, for example, in the way that such research is validated by numerous *citations* of previous results by other scientists.

This unadventurous policy is not necessarily dysfunctional. Science grows and slowly transforms itself by the accumulation and selective absorption of numerous valid contributions of small, but not insignificant, originality. The collective research of an invisible college, bound together by a network of literature, citations and social interactions, is an effective means for scientific progress.

It may appear however, that a major conceptual *revolution* is needed to make further progress. This requires some courage, since there is a tendency for the scientific community to repress scepticism concerning the basic principles of the current paradigm and to reject speculative efforts to replace it. But for some scientists this is a challenge – and an incentive to take risks in order to win the lavish reward for a highly original contribution that is eventually proved valid.

In either case, the *research programme* of the individual scientist is to a large degree socially determined, not only by the current state of knowledge in a particular field, but also by current opinions concerning significant *puzzles* and *problems* that are now capable of solution. The remarkable incidence of independent multiple discovery in science demonstrates this effect, as does the phenomenon of fashion in research topics and methods. This, in its turn, enhances competition for priority, and other communal factors already referred to, with further repercussions in the structure of the knowledge thus gained.

4.7 An over-idealized model

This model of academic science (Fig. 4) is highly schematized and idealized. No serious student of 'the science of science' should be encouraged to accept it just as it stands. The world would have little use for a body of knowledge generated by a closed community that disregarded all other social, cognitive, practical, or psychological considerations. The actual connections have profound effects on many aspects of the model. These linkages and interactions will be explored in the next chapter.

Nevertheless, within these very severe limitations, this model has a number of realistic features, and is not inconsistent with the way in which academic science works in practice. Thus, for example, it explains the personal commitment of the research worker to a discipline or field, the strong sense that scientific knowledge is objective and universally valid, the range and depth to which science is capable of exploring the natural world, the fragmentation and overall incoherence of scientific knowledge, the apparently progressive accumulation of knowledge, the self-sustaining dynamics of basic research, the wide range in the degree of certainty of scientific facts and theories, the co-existence of strict orthodoxy and speculative radicalism, the great importance of imaginative pioneers, the conventionality of science education – and many other familiar characteristics of science in its more basic modes.

At the same time, the social model provides no foundations for philosophical or ethical scientism. There is no scientific *method* giving privileged access to truth, and no scientific *attitude* deserving special respect as a virtue. The various objections to positivism are easily accommodated, without mutual contradiction and without recourse to complete scepticism or relativism. And although it is clear that any model of science that relies upon a separate scientific community is almost bound to be governed by a belief in science for its own sake, this is seen to be a functional ideology of this particular social group and not as an absolute moral imperative.

Conventional 'valid' science lacks a coherent image of itself. In an entirely uncritical manner, it conveys by implication a miscellaneous package of out-of-date materialist, positivist and technocratic views that can no longer be sustained philosophically or sociologically. Simply for its own good, to present a credible picture of its own operations, academic science needs something like this more naturalistic stereotype. At a time when it is under attack from many quarters, it would be wise to put its house in order.

This model makes no concessions to irrationality or antiscientific prejudice: on the contrary, it provides sound justification for the apparent intellectual anarchy, and competitive disorder that is often to be observed as science evolves. It shows, moreover, how individual freedom is an essential ingredient in the social system by which reliable knowledge is being generated, and makes a strong case for the autonomy of academic science as a whole.

But there, of course, we begin to touch upon more sensitive issues. It makes no sense to talk about academic science without reference to the benefits – actual and potential – that it can offer to human existence. These benefits are traded, within the social system, for material support for basic research. Without such external connections, forces and interactions, this model would seem to describe an esoteric ritual performed by a priestly sect. From the point of view of the non-scientist, it is precisely the social role of this special community that must be properly defined, and that is what we must now discuss. But it is impossible to bring to the surface the positive, creative aspects of this role unless one understands quite clearly how science really works within its own particular sphere.

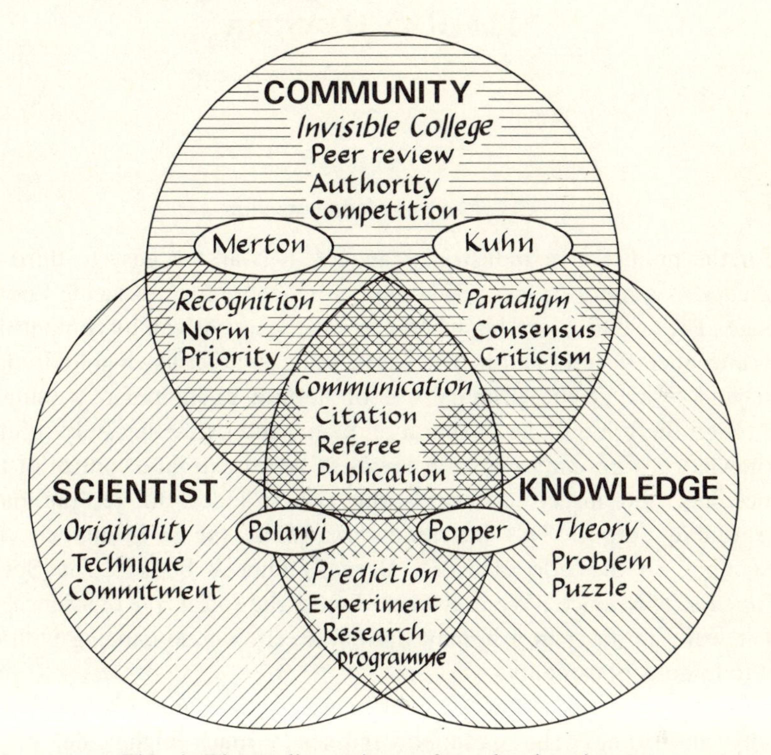

Fig. 4. Three dimensions of academic science

5

The R & D system

5.1 The knowledge machine

For the politician or industrial manager, it is all too easy to think of science as a more or less self-contained machine for producing knowledge. The scientists are big wheels or small cogs, driven by competitive or bureaucratic interaction. The whole thing is rather like a gold dredge, digging away at the primary ore of natural phenomena, passing it through the grinding and separating plant, and extracting fully refined knowledge. The main question seems to be how to take control of the machine, and direct it towards the richest lodes, to get the most profitable output. This instrumental attitude (§6.2) is implicit, for example, in the title of J. D. Bernal's famous book – 'The Social *Function* of Science'. We may even discuss the social *role* of science, as if it were an actor in a human drama, a self-conscious being with an autonomous personality.

But the image of science as a machine or an organism is fundamentally misleading. The metaphor implies a much higher degree of structural coherence and integration than is ever to be found in reality. It encourages big, bold, silly questions like 'How can science get us out of this mess?' or 'Is science a good thing?', for which there are only vain silly answers. If STS education is to make any progress, it must probe deeper than this into the complex web of social relations in which science is entangled.

One of the pioneering organizations in the STS movement calls itself SISCON – Science In a Social CONtext. This context, however, is multi-dimensional. As we saw in the previous chapter, even basic 'academic' science has psychological, sociological and philosophical attributes which must all be taken into account. The elementary social model of Fig. 4 is not just a collection of individual research workers, *or* a community bound together by social norms, *or* a body of organized knowledge: it has all these characteristics simultaneously. It relates to society in all these dimensions – through the scientist as a professional expert and as a citizen, through the formal institutions that shape the scientific community and sustain it within the body politic, and through the influence of scientific knowledge on human affairs and culture.

These relationships are not, of course, independent of one another. As in the elementary model of academic science, it is the interaction between the various components, across the dimensions in which they are defined, that drives the system along. But if we are to make any sense of such complex phenomena as technological innovation, or offer any opinion on such subtle issues as the social responsibility of the scientist, we must first disentangle these various threads of influence. STS teachers have widely different views on the exact interpretation of such phenomena and the proper response to such issues, but they are all in tacit agreement that 'Science', 'Technology', and 'Society' are related in many different ways, all of which may turn out to be significant.

In this chapter, therefore, the elementary model of basic science is enlarged and extended by linking its various elements to the society in which it is embedded. This reconstruction is still very schematic, and not to be taken literally. A number of highly controversial questions within the field of STS studies must be disregarded in the effort to arrive at a coherent description. Some of these questions will be discussed in chapter 6, but they will not be answered: it is not the purpose of this book to lay down a curriculum, complete with model answers, for STS education. But working our way round the extended model, drawing attention to the major connections and interactions, referring to the issues over which there is wide concern, we shall see that there is a framework within which almost all such issues find a place. This framework is no more than a very rough scheme. Different teachers and scholars would put very different weights of emphasis upon its various elements. But it is a representation of the place of science in society that should be acceptable in outline both to conventional science teachers and to the STS education movement.

5.2 Apparatus

The model of academic science outlined in chapter 4 obviously lacks one very important element: it entirely ignores the immense amount of *apparatus* that is needed to do research. In the past, the material facilities for experimental work were usually within the means of the individual scientist, or of the academic institution where he or she held a post. It was reasonable to regard them merely as personal instruments for the practice of this vocation. Nowadays, these facilities are always so expensive, and often so vast in scale that they constitute a separate factor in the research system. In addition to the psychological, sociological and philosophical dimensions of our model, we must appreciate that it has a significant magnitude in a purely material dimension.

This is as true of fundamental, academic research as of its technological applications. Particle accelerators, research reactors, space probes, radio telescopes, oceanographic research vessels, and such like instruments of *Big Science* cost millions of pounds at a time. They must be served by large numbers of technical staff, and are often used by many scientists simultaneously Expensive instrumentation is not confined to particular fields of research: for example, in almost all branches of modern biology, equipment costing tens of thousands of pounds is in routine use. The contemporary image of the scientist is projected against the background of a large laboratory building, crowded with apparatus and well supplied with computers, workshops, libraries and secretarial services.

The simple and obvious fact that modern science is no longer entirely labour intensive, but requires a substantial technical infrastructure and heavy capital investment, is of the highest significance for our theme. On the one hand, it has a profound effect on the internal functioning of the science system. Thus, for example, the research programme of the individual scientist (§4.6) may be determined less by the urgency of a problem requiring solution than by existence of a piece of apparatus to which he or she has access. The competition for recognition within a particular field of research may be significantly perturbed by the availability of the necessary facilities. Because of the size and complexity of these facilities a number of independent research workers may be forced to co-operate in a *research team*, which thus, so to speak, takes the place of the 'individual scientist' of the schematic model (§4.3). In such a case, however, it is difficult to ensure that the recognition and other rewards for successful research are not unfairly allotted to the leader of the team, thus further enhancing the stratification of status and authority within the scientific community (§4.4). These are all quite real consequences of the increasing importance of the apparatus and other material factors in scientific research.

On the other hand, academic scientists can no longer find the means of financing the apparatus they need from their ordinary personal and institutional budgets. Support for research facilities must come on a large scale from outside the basic science system. Here is one of the major links between 'science' and 'society'.

As we have seen, academic science has the capability of being intellectually, socially, and psychologically self-sustaining. It can provide the members of the scientific community with programmes for research, and incentives to carry them out, without apparent reference to external demands or needs. But it cannot simply wind itself along in this way unless it is well provided with material resources. This is the logic of the situation.

In principle, therefore, research can no longer be an autonomous social activity, carried out by a community of scientists of their own free will. It must always be heavily subsidized by the major organs of society at large, to the extent that it could scarcely exist without such subsidy. This support is usually justified by reference to the benefits that may be expected as an outcome of the research, whether of practical or abstract knowledge (§6.2). In the long run, the apparatus and other facilities for research are supplied, not to gratify the intellectual aspirations of the professional scientists, but to further the ends of the state, or of industry, or (ideally) of the whole community. This again is a point on which STS education must necessarily insist, against what is implied in conventional science teaching, where valid science is often regarded as an end in itself, worthy of support in its own right.

5.3 Technology

The distinction between 'Science' and 'Technology' is very vague. There are no sharp discontinuities as we go from the most academic pure research, through applied science, to industrial development and technical innovation. Knowledge is drawn from the same archives, experiments are done with the same sort of apparatus, by people with the same education (§5.8) and the same expert skills (§6.3). Since our whole civilization is permeated with advanced technology, this is the great open frontier between science and society. Science connects with society, and flows into ordinary life, through technology.

But technology is not merely the material product of science. Whether we regard it narrowly as knowledge about practical techniques, or more broadly as all that is actually created by such techniques, technology is as old as human culture, and has both roots and branches right outside the realms of experimental and theoretical research. Indeed, because of the immensity of technology in the economic, political and cultural dimensions of society, it is easy to regard science itself as a mere appendage or instrument of this dynamic force – like the research laboratory of an industrial corporation, dwarfed by the factory buildings and offices where the real work of making and selling is done. Impressed by the power of technology in peace and in war, we readily fall into the instrumental attitude that regards science simply as a machine for the solution of technological problems or for the automatic generation of new technological devices.

This attitude is dangerously simplistic. Carried to its logical conclusion, it suggests that the whole question of the social context of science can be displaced towards a discussion of the impact of technological innovation on the quality of life. By this line of reasoning, STS education has come to be associated with a series of issues, such as environmental pollution and energy conservation, which are really much broader and more contentious than can be fairly dealt with in an educational curriculum (§6.5).

The importance of these issues is not to be disputed. There is much that needs to be said about the social dynamics of technology, its incentives and imperatives, its capabilities and limitations. The exclusion of all such questions from conventional science education carries implications in favour of uncontrolled technological growth and change that deserve serious criticism. But it is equally uncritical to conflate 'science' with 'technology' and 'technology' with 'industrial civilization', and then to attribute to 'science' a catalogue of ills and vices that is no more real than the conventional list of benefits and virtues (§7.8). A serious attack on these questions leads inevitably to a study of the political and economic foundations of modern society, and hence to the great and abiding themes of public affairs – capital and labour, socialism and free enterprise, private affluence and public squalor, charity and self-reliance, corporations, nations and people, government and anarchy, class, power, freedom, violence, legality, and the rest. Not only are these themes so controversial that they lie outside the permitted boundaries of deliberate instruction in a pluralistic society: they are just too big and important to be dealt with under the heading of STS education. There is no justification for turning this subject into a vest-pocket course on political and economic theory – a course that is liable to be as naively antiscientistic as the scientism it is intended to combat.

Nevertheless, STS education must concern itself in some depth with the social relations of technology. As we saw in chapter 1, the great majority of students of academic science eventually find employment in various technological professions, as technicians, engineers, physicians, industrial managers, etc. It is essential that they acquire some general grasp of the social implications of any such career, and the political and economic conditions within which it is to be followed. In detail this preparation must be part of the technical training for each particular profession – for example, as part of the engineering or medical curriculum (§7.3). But it is the responsibility of STS education to build up a general understanding of 'the place of science and technology in society', as a foundation for the more specialized and diverse conditions to be encountered in different technological careers.

In the end, the treatment of those aspects of the uses of technology that touch directly upon controversial political or ethical values must be entrusted to the wisdom and professional integrity of the STS teacher. Some of the ways in which these topics can be dealt with in practice will be discussed in chapter 7. The formal STS curriculum must concentrate on a core of facts and opinions on which STS teachers largely agree. This area of consensus is by no means trivial, or immediately obvious to the average science student. At this point, therefore, we return to our simplified model of academic science, and show how it is connected with, and modified by, science-based technology.

5.4 Technology as knowledge

The word 'technology' is now used rather loosely for the whole of the technical activity of a human culture – the industrial products, the machines, the vehicles, the factories, and all. Strictly speaking, however, it means the body of organized *knowledge* about such techniques. That is to say, it defines itself alongside science, in the knowledge dimension. Superficially, scientific and technological knowledge appear so very similar that it seems pedantic to differentiate them. But despite a great deal of overlap, and a continuous gradation of characteristics from one to another, not all technological information is based on science, and not all science is derived from technology. This is obvious if we think of extreme types – high energy physics and molecular biology at one end, for example, and mechanical engineering and agriculture at the other. There are significant differences of ends and means which are important factors in the STS relationship.

Science differs profoundly from technology in the purposes for which knowledge is accumulated, and hence in the criteria by which particular items are chosen for inclusion. The objective of academic science is to create the widest, most complete body of consensual knowledge, without particular regard for its use. Problems are chosen, research is undertaken and the results appraised according to a variety of principles, such as 'explanatory power', 'intellectual consistency', 'descriptive accuracy', and so on, which make no reference to the practical benefits which might arise from them. Even if such benefits can be perceived in principle, the knowledge is made available as a *public* resource, in an organized form.

Technology, on the other hand, is essentially instrumental in its valuation of knowledge. Information is accumulated and treasured for the sake of what can be done with it. The goal of technology is to preserve and improve the techniques to which it relates – the production of food and material goods, the use of weapons and tools, the preservation of health, the establishment of communications, etc. To put it crudely – technology is just 'know-how', writ large.

This distinction is obviously very schematic. But it explains why technology is much more loosely organized than science in this dimension. Technological knowledge is often generated simply as an intermediate step in the creation of a new technical process, within a particular industrial corporation. It is not regarded as having value just for itself, and there is no imperative to make it public at all. Thus, even if it is not actually kept secret, it is seldom made fully explicit in publishable form, ready for inclusion in a catalogued archive. In any case, technology is not validated by discussion and criticism: it proves itself by its success in practice, and has no particular occasion to attain the standards of inherent plausibility, observational reproducibility and logical consistency which are demanded of scientific communications. However effective it may turn out to be in the real business of making and doing, technology appears much less coherent and accessible than science as a body of knowledge.

Nevertheless, scientific and technological knowledge turn out to have a lot in common. Much knowledge that was originally acquired 'scientifically', out of sheer curiosity, without any practical use in mind, has proved later to be extraordinarily useful. We have only to think of the discoveries of Faraday, Maxwell, and Hertz, concerning electromagnetism, which found quite unforeseen application as the new technique of radio communication. When this happens, the original basic science is automatically incorporated in the knowledge system of the new *science-based technology*. This historical phenomenon is now so familiar that we tend to regard all scientific knowledge as of potential technological value. Who would now discount the future applications of molecular biology in agriculture, or can confidently assert that mechanical engineering will *never* benefit from what is learnt in high energy physics?

Looking at the relationship from the other side, we note a tendency to turn every useful craft into a *scientific technology*. As technical experience is recorded and codified, it becomes suitable material for academic study. In due course, this generates further research, to extend and organize the subject theoretically. Eventually, the original technological goals are given second place to the accumulation of knowledge, apparently for its own sake. Clinical medicine, for example, was a practical craft in which a great deal was learnt about health and disease; from this learning developed the basic sciences of anatomy, physiology and pathology, whose primary objective is a thorough understanding of human biology, regardless of clinical application. But this proliferation of academic effort in the direction of fundamental knowledge does not leave the original technology unchanged. A theoretical framework is now available, within which therapeutic techniques or design principles can be rationally interpreted and justified to practitioners. This historical process, also, is so familiar that every technology now seeks to advance itself into a new branch of science.

By one means or another, the scientific and technological knowledge systems tend more and more to overlap. It matters little whether an advanced technology grew originally out of a novel scientific discovery, or whether it became more scientific in the course of its natural development from craft beginnings: in either case, its whole professional practice is permeated with concepts, theories, data and techniques that are also widespread in the public archives of academic science. A great deal of 'valid science' is an indispensable component of the knowledge required in engineering, medicine and similar technological professions; conversely, the technical literature of these professions is intelligible to, and significant for, the academic research worker in basic science.

In our contemporary civilization, the interface between science and technology has been almost obliterated. But even where they overlap and interpenetrate, the two systems of knowledge are seldom identical in form and content. On the scientific side, delicate webs of theory are spun to trap vast stores of information, out of sheer curiosity, without any plausible practical motive. On the technological side, empirical data, designs, and techniques, are accumulated in great quantities, with little reference to their underlying significance. Much of what the technologists know seems inexplicable by scientific theory; much of what concerns the scientists is quite incapable of application. What seems important in constructing an ordered pattern of theory may be quite subsidiary to what is needed to build a bridge or transplant a kidney. And there still remain very distinct and 'pure' sciences, such as astrophysics, which are pursued with little thought of practical gain, just as there are still crafts, such as pottery, which have not moved far into the scientific era, and which owe much more to trial-and-error experimentation than to deep conceptual analysis of characteristic problems.

5.5 'R & D'

Technology never stands still. In itself, and in its interactions with society at large, it is a dynamic force. Of all topics in STS education, there is none more challenging than the question of the sources of *technological innovation*. It is the continual change in the techniques, the products, the processes, the capabilities, the costs, the benefits, of advanced industrial civilization that puts so much strain on our political and economic judgements.

One such source has already been noted – the exploitation of the possibilities of a new scientific discovery. Another source of technological innovation is equally unpredictable but more traditional – the invention of novel processes, machines or products within the framework of techniques and principles of an existing technology. It is often impossible to distinguish sharply between these two sources: there may be a scientific novelty at the heart of a successful invention, and an act of high imagination is often required to envisage the practical possibilities of a new scientific idea. Technological innovations arise from a wide variety of sources, ranging from deep scientific theory (for example, the laser) to hard practical experience (for example, most components of an automobile).

As a technology becomes more sophisticated, however, it is forced to adopt a more deliberate and rational procedure for change. It is characteristic of an advanced technology, such as telecommunication engineering, that it can no longer maintain a satisfactory rate of innovation by relying on a random supply of independent inventions. Even without excess pressure of commercial or military competition, numerous practical problems arise in exploiting a novel invention, or taking advantage of a novel idea to improve an existing product. These problems are often so interlinked that they cannot be solved on an *ad hoc* basis as they arise, but must be foreseen and solved in advance.

Every major industrial corporation involved in scientific technology is thus forced to support *technological research and development*. Its future survival and prosperity cannot be assured without an adequate supply of new design concepts, new manufacturing methods, or other novel procedures. In the same spirit, government agencies responsible for health, agriculture, environmental protection, defence, transport, etc., undertake 'R & D' on a large scale to solve their multifarious problems.

In many respects, 'R & D' is not so very different from research in academic science. It uses the same types of apparatus, the same paradigmatic frameworks of fundamental concepts, the same intellectual techniques of experiment and theory, speculation and observation, data collection and hypothesis testing. To the extent that all research calls for the solution of a series of relatively novel technical problems, academic research in a basic science is practically identical with *strategic research* – very long term investigations whose ultimate applicability is only vaguely conceived. As the problems to be solved approach the realities of production or use – that is, at the final stage of *development* of a novel device or process – there is a shift in emphasis towards short-term, imperfect, expedient solutions to meet a very specialized range of technical criteria. But there is still the spirit of experimentation in the work: a pilot plant or prototype aircraft under test is rather like a piece of apparatus whose satisfactory performance confirms the theoretical principles on which it was designed.

Nevertheless, there is a fundamental distinction in principle between 'R & D' and our academic model of scientific research. In this model, each scientist is personally responsible for a programme of research (§4.6) which is not laid down by an external authority. Since this programme is designed to achieve recognition for its contribution to knowledge, it is not entirely arbitrary, and is strongly influenced by various forces within the scientific community. But there is no overall goal for the evolution of scientific knowledge, and no collective plan to follow particular lines of research to reach particular objectives.

The essence of 'R & D', on the other hand, is that it is *mission-oriented*. It is programmed to support the policies of a specific organization – the investment and production plans of a commercial corporation, the battlefield scenarios of the military, the health or welfare benefits desired of a government department. The ultimate objectives of such a programme are not within the control of the scientists who carry out the research, but are laid down by powers outside the science/technology system. In the shorter or longer term, in a narrower or broader sense, the criterion by which 'R & D' must be judged worthy of support, and 'R & D' workers deserving of reward, is the contribution that has been made to the success of the mission towards which it is oriented.

Here again, the boundary between 'science' and 'technology' is not nearly so sharp in practice as it seems in principle. By definition the outcome of all research is unpredictable. A programme for research or technical development is not like an engineering blueprint. Methods and objectives must be continually revised as new results come in. This revision can only be undertaken by those who understand what is going on – that is, mainly by the 'R & D' workers involved in the project. In principle, a scientist engaged in 'R & D' is no more than a professional expert serving the interests of the corporation by whom he is employed; in practice, he may have as much discretion in the choice of research methods and the evaluation of the results as if he were a 'self-winding' academic scientist.

In any case the personal freedom and autonomy of the 'pure' scientist is exaggerated in the academic model. A great deal of basic science is already concentrated into research teams (§5.2), whose members are strongly constrained to align themselves with the programmes, policies and objectives of the group. The options open to the individual research scientist are severely limited by his personal experience, the 'state of the art', the availability of resources, and the range of potentially soluble problems suggested by the current paradigms (§4.6) of his subject. These options are not necessarily made much narrower by an instruction to follow some very general research policy, such as seeking an understanding of metal fatigue or of schizophrenia.

Taking a broader view, the academic scientists involved in basic research can be regarded as participants in the overall 'R & D' effort of society. They are given the research facilities and the freedom to choose their own research programme, within limits decided by the community of their scientific peers (§4.4), because society needs the discoveries they will then probably make. Experience has shown that this is a very effective means of meeting the very long-term goals of society, in the form of a public archive of reliable knowledge (§4.5) to be drawn upon for all manner of future benefits.

5.6 Research management

The elementary model of academic science says nothing about how a scientist should earn his daily bread. There is almost a taboo against any suggestion that research is a way of making a living. It is just taken for granted that any scientist worthy of recognition has some means of support.

This convention is maintained in *academia* by the polite fiction that research workers are officially employed as teachers (§5.8). In fact, the whole career structure of higher education is adapted to providing a suitably graded range of temporary and permanent posts for scientists and other scholars of the appropriate degrees of distinction. The most tangible reward for an original contribution to science is an appointment as a lecturer or professor in an academic institution.

Since the curricula of higher education are supposed to mirror the paradigms of contemporary knowledge, this arrangement works quite well, especially in the basic scientific disciplines. It allows the academic scientist to go about research without strong pressures from society at large. The delicate web of norms and sanctions within the scientific community (§4.4) is also protected from the stresses that could arise if research performance could be demanded specifically under a contract of employment, or could be directly traded for cash. Indeed, academic scientists whose apparatus is paid for by a research grant from an external agency such as a government research council are also shielded from pressures arising inside the institution by which they are employed.

An established academic scientist may thus enjoy a remarkable degree of independence in his or her investigations. Although the research may call for quite considerable resources of equipment and technical assistance, it is managed only very indirectly by committees of peers in a grant-awarding agency. As academic science becomes more closely incorporated in the general R & D effort of society, there is a tendency to make research a full-time occupation, dissociated from higher education. Even in such fundamental disciplines as molecular biology, the demand for knowledge is felt to be so urgent that only completely dedicated research scientists, working together in large institutes, could satisfy the need. But every effort is made, in the organization of such institutes, to leave established scientists as free as possible to 'do their own work'.

Nevertheless, this personal independence is not given nowadays to most scientists and R & D technologists, who are mainly employed as full-time professional workers, in large organizations, served by sophisticated and expensive apparatus (§5.2). Economic logic speaks for the efficient utilization of research facilities by hiring research workers with particular skills to perform particular functions in research teams devoted to specific projects. In Big Science, it is difficult to oppose the rationality of a stratified administrative structure, in which individual scientists are made subordinate to the higher echelons. Despite the desire to preserve the academic style of research, the management of a large laboratory cannot be so egalitarian or participatory as the conference of a scientific invisible college (§4.4).

This tendency becomes much more pronounced as one moves into more technological fields. The development of a new industrial product or a new weapon is no game for the enthusiastic amateur. It involves closely co-ordinated action by large teams of skilled personnel, working against the clock under firm direction. A great deal of R & D is motivated by patriotic or commercial rivalry. It is driven by conflict between national government or industrial corporations, not by competition for recognition between individual scientists. By the logic of the situation, this sort of work can only be done by full-time employees, giving their services and obeying orders in return for payment.

Organized R & D thus induces a radical transformation of the social relations between scientists. The personal commitment of the individual scientist to the advancement of knowledge gives way to loyalty to the project, or laboratory, or company, or country for which he or she works. The norms and sanctions of the scientific community yield place to managerial responsibilities and directives. The traditional incentive of recognition for published research is replaced by the more tangible reward of a pay rise, or promotion, for service beneficial to one's employers. The social role of the scientist as an employee of a bureaucratic mission-oriented corporation is not really consistent with the traditional career of an independent member of a loosely structured, universal, scientific community.

In the more academic, more public activities of science, this contradiction is not felt so sharply as in industrial and military R & D. But it is inherent in the whole concept of the *management* of research, which follows naturally on the incorporation of science into the technological, industrial, and governmental organs of modern society. In a general way, we may talk of all science having been *industrialized*, even in its most basic and academic branches.

5.7 Science policy

A significant proportion of the material resources of a modern industrial nation now goes to scientific and technological R & D. The allocation of one or two per cent of the national income to this intangible purpose calls for administrative procedures by which science is related to larger political and social goals.

The calculation of total R & D expenditures is not very meaningful, except as a general indicator of the scientific 'strength' of a nation. It would include, for example, the very large sums that must be spent on the development of new industrial products, according to budgets that have already been decided from other considerations such as the expectation of profit. Thus, for example, a project for a new design of nuclear-powered submarine, at an overall cost of X billion dollars, would include provision for sums of the order of X/n billion dollars (where n is a number of the order of 5) to flow over a period of many years into scientific laboratories and engineering test facilities. The R & D must be 'bought' as part of the project, along with the steel and the shipyards, and 'managed' wisely to obtain an optimum return. The same would apply on a more modest scale to the introduction of a new drug or a new type of ball-point pen.

Investigations and experiments of a scientific type are thus undertaken by a whole range of independent organizations, in industry and government, to support their normal ongoing activities. Since these activities satisfy diverse social and economic interests – commercial profit, health, environmental protection, defence, etc. – they bring scientific and technological factors into all the political issues of the day. But this involvement is usually indirect: the management of the relevant R & D does not take precedence over the economic, health, environmental or defence policies of which it is part.

But for more strategic research, whose prospective benefits are less certain and more distant, administrative dependency on the normal activities of industrial corporations and government agencies is often unsatisfactory: long-term potentialities are easily sacrificed to short term needs. For the furtherance of fundamental disciplines whose applications cannot be envisaged, there is practically no substitute for government patronage. 'Academic' science may be no more than 10–20 per cent of all R & D, but it still amounts to a large sum that can easily be discerned in the accounts of the nation. By this cash nexus, science is linked directly with society at the governmental level.

The primary function of *science policy* is to make a rational allocation of central government funds to various fields of research. Since basic science is almost totally dependent on government funding, through one channel or another, this is not just marginal support by an occasional judicious subsidy. The government is buying basic science, as a long-term resource, for the good of the nation.

This is an activity that cannot be governed by precise calculations of costs and benefits. The *economics* of R & D may say something about general criteria for successful investment in research, but gives very little guidance as to the proper sum to be spent on a particular type of research, at a particular stage of maturity of knowledge, possibly relevant to particular social goals. The rational assessment of ends and means is as much an art in this branch of public administration as in other fields of politics.

In making such decisions, the opinions of lay persons, such as politicians, administrators and accountants must be given adequate weight. This is not always easy, since the judgement of experts on the technical possibilities of a programme of research cannot be disputed. That is to say, any plan for research in a basic science must be consistent with the current paradigms in that science (§4.6), as they might be expressed, for example, by the leading authorities in relevant fields of knowledge.

For this reason, the policy-making process must not be ill-matched to the social structure of the corresponding scientific community (§4.4). At the microsocial level, this is achieved by a peer review procedure in which grants are made to individual academic scientists for particular research proposals of 'timeliness and promise'. Since the members of review panels are chosen from the corresponding invisible colleges of scientific colleagues, this puts each individual scientist a little more under the thumb of communal opinion than he would prefer, but leaves the social relations of the simple model of academic science in good working order.

But this procedure, in which the academic scientific community is left to share out the money according to its own internal criteria, is unsatisfactory at higher levels. The traditional scientific community has no institutional structures for deciding, say, the relative allocation of funds between major scientific disciplines, or the scientific value to be attached to a large single project such as a particle accelerator. Its traditional leaders are authorities in their own fields of scholarship, but no wiser than many laymen over wider ranges of subjects and issues.

Through a variety of administrative devices, such as advisory posts and research councils, they play an essential role as intermediaries between scientific and governmental circles; but science policy in general is now too important to be decided entirely by the scientists. Basic science has moved from the periphery towards the centre of society, and is thus brought under the control of the state. The scientific community described in the simple academic model of chapter 4 continues to enjoy considerable internal autonomy as a distinct 'estate' of society, but only within the context of government decisions on quite specific issues such as the initiation of large research projects or support for socially relevant scientific themes.

5.8 The educational interface

Academic science is obviously an essential part of the educational system. Many research scientists are employed as teachers in higher education (§5.6), where they are brought into contact with a much wider range of students than those destined for scientific careers. As we saw in chapter 2, the science curriculum at all levels must satisfy standards of validity spreading downward and outward from the high science of the research frontier. As this book seeks to demonstrate, science education itself is a major interface between science and society, through which flows a powerful current of imagery and ideology.

The R & D system has a fundamental stake in education, simply to supply itself with well-qualified scientific and technical workers. For its purpose it is not sufficient to rely on the individual initiatives of young people fascinated by the romance of science as a career. Mental ability and emotional commitment must be moulded by formal instruction in valid science, from the middle school onwards. As science moves closer to the political and economic hub of society, it acquires more and more leverage and more and more responsibility in the educational sphere. High scientific talent is rare; it must be identified at the earliest possible moment, and trained efficiently to fill its important social role. Sophisticated research apparatus (§5.2) must be tended by correspondingly sophisticated technical staff. The long-term supply of skilled personnel has thus become as significant a factor in science policy as the provision of financial resources for research.

Nevertheless, the connections between research and education are not very direct. The simple model of academic science does not have science education as a primary factor. It takes for granted a ready supply of would-be scientists (§4.3) seeking recognition, but says nothing about how the new cohorts should be recruited and trained. The norms of the scientific community (§4.4) do not rule over the transmission of knowledge from generation to generation, nor do they deal with the delicate social relationship between teacher and pupil. The philosophy of science (§4.5) almost completely ignores the central paradox of science teaching – the fundamental contradiction between sceptical and dogmatic attitudes towards 'established' knowledge.

At a certain moment, of course – typically, entry to a *graduate school* as a *research student* – those who are sufficiently talented are admitted as apprentices into the scientific community. Henceforth, they are, regarded as junior members of that system, seeking recognition by original yet valid contributions to knowledge, bound by the norms of communality, etc., and required to become thoroughly acquainted with the knowledge relevant to their specialty. Graduate courses, for example, are mainly directed towards the diffusion of new concepts within the scientific community itself, and have little significance outside the world of advanced research.

Technological research and development, by its very nature, is separated organizationally from higher education. The necessities of formal certification for the practising doctor or engineer impose curricular constraints and standards on technological education (§1.3) which are partially transmitted further down the system to school science. But university medical and engineering faculties are just as remote and 'academic' in their attitude to science education as their scientific colleagues.

The *concern* of academic science for the contents, methods and objectives of science education is thus much more limited than its obvious *influence* would suggest. The forces that might be expected to flow, from society at large, inwards through the system of secondary and tertiary education, are impeded by institutional stratification and the sharp demarcation of the professional roles of the school teacher and research scientist. Despite its immense importance in an advanced technological civilization, science education is not closely integrated within the R & D complex, and is only loosely bound to science itself.

5.9 Common knowledge of science

Above all, science is a body of knowledge. It encompasses vast amounts of information, organized around a variety of theories, conceptual schemes, categorial frameworks and other interpretative principles. This knowledge is a major component of the culture of modern society, for whose benefit it has been accumulated, and to whom, in the last analysis, it belongs.

Much of this knowledge is eventually used by society indirectly through its technological applications. But there is also a very important interface between science and society in the cognitive dimension. That is to say, there is a direct connection between what is 'known to science', and what is 'common knowledge' in society at large.

Of course, there is an immense range and diversity in what people know or believe on scientific matters (§1.5). In a very broad sense, however, we can talk of 'scientific world pictures' in which might be found the major theoretical schemes of the basic scientific disciplines (§3.2), very crudely summarized and roughly fitted together into a makeshift representation of macroscopic and microscopic nature, the earth and the heavens, molecules and materials, cells and organisms, forces and fields. From such representations we each of us make various metaphysical and practical deductions that help us along the way through life. In other words, science both contributes to, and is a part of, the *ideology* of contemporary society (§6.5).

Modern civilization is also 'scientific' in a narrower sense. Everybody becomes acquainted with a great many specific items of information derived from science – what foods are health-giving, how minor ailments should be treated, what happens when a fuse blows, how to make a garden grow, and so on. This information is usually fragmentary and untheoretical, since it is related to the immediate circumstances of everyday life, in work and in play. By the standards of valid science it is often very incomplete and inaccurate. Nevertheless, it is upon such general and specific conceptions or misconceptions of scientific reality that a great deal of social action is ultimately based.

The discrepancy between scientific knowledge and folk knowledge is notorious. Yet science, by definition, is public and open (§3.4). Although the scientific archives are not usually housed in municipal libraries, they are open for study in universities and reference libraries for almost any earnest enquirer. How is it that science is, in fact, a closed book to most people?

Scientific knowledge is not secret, but it is decidedly esoteric (§3.4). In primary form, it is spread through hundreds of thousands of separate communications, each addressed to the research workers in a very specialized field. By reason of their complex nomenclature and technical phraseology, the vast foundations of previous knowledge they take for granted, and their profusion of observational and theoretical detail, such communications are almost impenetrably obscure to the layman. This applies as much to the scientist in another field – for example, the physiologist seeking knowledge about quantum chemistry – as to any member of the general public.

In this respect, basic academic science is no worse than the rest of the R & D system. Indeed the technological information that might be gleaned from blueprints, test data, instruction manuals, or direct observation of industrial practice, is often deliberately concealed, in the name of proprietary or military security. But even when it is not secret, scientific and technological information does not communicate itself freely except to those who are already knowledgeable on the subject. The purely mechanical problem of discovering what is actually known about some particular scientific or technological question often calls for an elaborate computer program to 'search the literature' of the subject.

Academic science requires for its own purposes a system of secondary services and publications in which the primary literature is periodically reviewed and consolidated, mainly for the benefit of the relevant invisible colleges. Through review articles, conference talks, symposium volumes and monographic treatises, the results of research are catalogued, criticized, and refined into established scientific knowledge. This paradigmatic material is incorporated, in simplified versions, into textbooks and thus transferred, through the educational system, to a wider public. The academic community regards itself as responsible for the validity of the knowledge that flows out thus into the common knowledge of society.

This pedagogic channel, although wide and deep, runs very slowly. It may take twenty or thirty years for an important scientific development to find a secure place in the school curriculum. The knowledge of science that most people have got from their schooling is usually at least a generation out of date. Recent science is picked up much more haphazardly through the mass media or by private reading.

The scientific community does not take any initiatives, or responsibilities, for the *popularization* of science. It is left to scientists to respond individually to the demands and temptations of the media, where there is always a market for interesting, well-expressed information on contemporary scientific developments. That is to say, scientific knowledge is 'pulled' into the public eye by popular curiosity and concern rather than being 'pushed' out into the world by the R & D system. In any case, there is seldom direct and spontaneous communication from the pen or voice of the research scientist to the eye or ear of the citizen. The content and quality of what is transferred across this interface depends crucially on the ability of professional science writers, who select, combine and transform valid scientific knowledge into books, newspaper articles, radio broadcasts and TV programmes for the general public.

Common knowledge of science is not under the control of the scientific community, with all its norms of universality and disinterestedness, originality and scepticism (§4.4). The extent to which it is coherent and reliable is not subject to strict logical criteria, but depends largely upon the standards of intellectual integrity in the organs of mass communication through which this knowledge flows. It also depends upon the social role of science and of the R & D system of which it is part. It seems that science does not fit as smoothly and neatly into its cultural context as most scientists and their fellow citizens would like to be the case. Nor, as Fig. 5 illustrates (p. 88), is research and development quite such a simple and direct social activity as people are often led to believe.

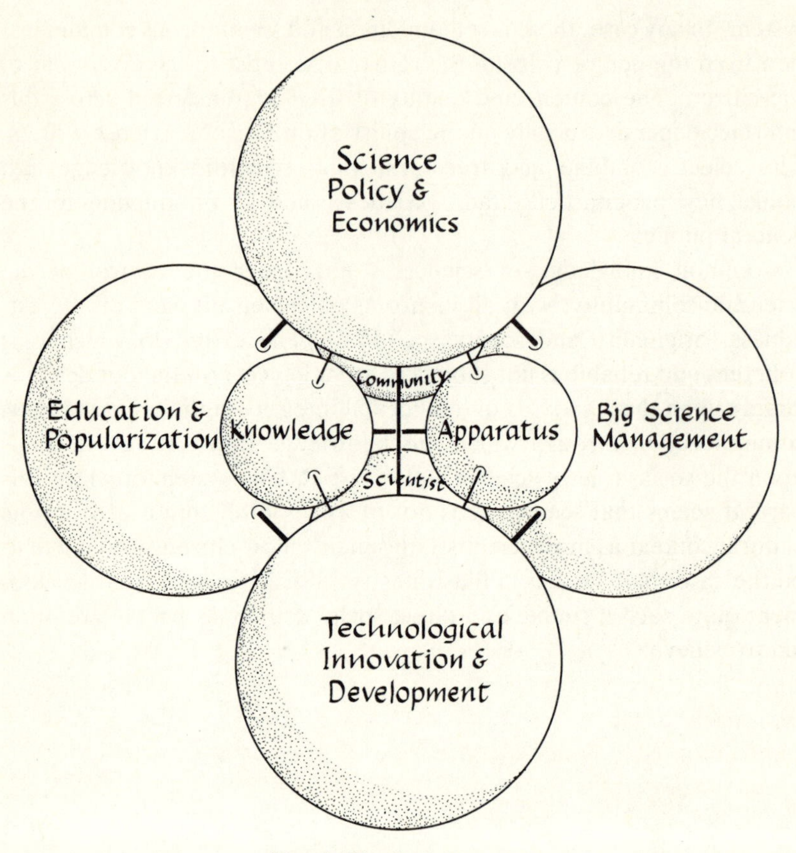

Fig. 5. The R & D system

6

Controversial STS issues

6.1 The underlying consensus

The last two chapters have been rather abstract. There has been practically no reference to empirical facts, and the line of argument always seemed to stop short of the really interesting questions. What is the point of outlining a purely schematic model of the R & D system?

The very possibility of there being general agreement on such a representation of the social context of science is fundamental for the future of STS education. It is this unacknowledged consensus on the underlying structure of the subject that can bring coherence to the curriculum. This particular model is not necessarily the best possible, but it was worth describing in outline, not because it is sharply disputed, but because there is so little realization of how much, in fact, is implicitly held in common by those involved in the STS movement.

This underlying structure is somewhat more complicated than scientist philosophers or sociologists used to believe. But it cannot be made simpler without grave distortion. To make sense of 'academic' science, for example, one must take seriously all three aspects – the *psychology* of research and invention, the *sociology* of the scientific community, and the *philosophical* criteria for objective knowledge. The immense resources needed nowadays for scientific *apparatus* cannot be ignored. In a wider social context, science is seen to be inseparable from *technology*, making up an *R & D system* whose internal *management* and public *policies* have profound effects on science itself and in society as a whole. Scientific and technological knowledge also has direct *educational* and *cultural* significance in modern civilization.

All these aspects of science are clearly relevant to the STS theme. Not everybody would agree on the weight to be assigned to each component of the model, nor on the exact manner in which they are related to one another – but I believe that this is how most STS scholars now picture modern science and technology.

It would be a serious mistake, however, to design the STS curriculum solely within the pallid scholarly constraint of only teaching those things about science on which everybody largely agrees. The basic facts about science and technology must always be kept in mind and explained clearly by STS teachers; but the subject will never come alive unless it opens out into more controversial issues, where facts and opinions are much more questionable.

The model itself needs much closer scrutiny in many of its details. Serious scholars of the sociology and philosophy of science, for example, could point to the many discrepancies and contradictions that arise in the detailed interpretation of such concepts as Kuhnian paradigms (§4.6) or the norms of the scientific community (§4.4). But questions of this kind are very subtle (§§7.7, 7.8) and may be somewhat too academic for STS education in general. The same criticism would apply to the innumerable questions that could be raised concerning, say, the economic benefits of R & D, or the management of scientific, technological, and educational institutions in various countries.

There are, however, a number of highly controversial and socially significant issues that arise, so to speak, around the edges of the R & D system. For most of us, the real concern about science and technology is whether they fit properly into their general social milieu, whether they do what people want of them, in the shorter or longer term, and how their professional participants – academics, research scientists, and other technical experts – should respond to the complex social roles into which they are often thrust. As we saw in chapter 3, conventional science education conveys characteristically scientistic responses to these questions. But scientists themselves do not see the world in that way (§3.9). In the present chapter we can look into such matters much more deeply, guided by our much more realistic model of the research process.

6.2 Science as an instrument

Most people are concerned that science should be properly *used*. In other words, they think of it as the *means* by which certain desirable *ends* might be reached. This attitude is implicit in the very notion that there should be a 'policy' for science.

Taken too literally, the analogy of science with an instrument is dangerously misleading. As we have seen (§5.1), the R & D 'system' is a loosely organized social activity which cannot properly be compared with a machine, or sent into action like an army. Nevertheless, the deliberate *planning* of research on particular topics, with the expectation of arriving at results with preconceived applications, is a characteristic of modern society (§5.7). It is not far from the truth to say that society wants to buy scientific and technological research in much the same spirit as a householder might buy a hammer or an electric toaster.

But what style, or make, of research should be bought? Many serious social critics have proposed lists of 'ends' to which science ought to be directed, or have drawn attention to questions of great social significance that have been neglected by science. This is one of the major themes in STS education – and one of the most difficult for the student to get into perspective.

The trouble is that the ultimate goals of research could be as diverse and multifarious as our most fantastic dreams. The ancient alchemist sought an Elixir of Life, for immortality, a Philosopher's Stone, for infinite wealth, and a Perpetuum Mobile, for endless leisure; his imagination was no wilder than our present-day search for a cure for cancer, for the materials to sustain endless economic growth, and for fusion power from sea water. It is easy to make fun of sci-fi objectives – until one recalls the actualities of antibiotics, nuclear weapons, and instruments landed on Mars. In the light of recent history, the range of possibilities that might be opened up by well-directed research is far wider than the most vivid imagination can conceive.

Almost all these possibilities, however, are entirely conjectural. The attainment of any one of the ends that we might now desire is seldom more than a wild hypothesis. It cannot be made a firm anchor, on which thick strands of policy could be allowed to depend. A catalogue of such items is useful primarily as a checklist to recall to mind those very few projects that are to some small degree capable of realization.

The objectives of research are selected much more by whether there is the least prospect of making progress towards them than by carefully weighting up their intrinsic merit. It is only when there seems a plausible 'means' of getting to it that a desirable 'end' can figure seriously in a policy or plan for the application of science to human needs. This limitation of the instrumental attitude is not always perfectly evident to those who look at modern science and technology superficially from the outside.

The large proportion of sophisticated R & D that is devoted to the development or modification of existing techniques gives a misleading impression of what can be done by brute force. In the short run, a major R & D effort may be relied on to reach limited goals that are already envisaged as an extrapolation of the current 'state of that art'. This is very obvious in military projects, where cost is no objection. But the apparently inexorable progress of certain industries, such as electronics or aviation, would not necessarily be repeated in other fields simply by using R & D in the same immense quantities. The disappointing outcome of the 'War on Cancer' is a case in point. The history of technical progress is often gravely misunderstood. With the benefit of hindsight, our attention is caught by policies and predictions that proved to lead along the correct path, and we tend to overlook the many blind alleys that may have had to be explored on the way. Or we may be so blinded with science that we do not realize the extent to which technological innovation arises simply by the accumulation of practical experience, and has had to adapt its uses, by trial and error, to the capabilities thus discovered.

The problem of using science properly thus penetrates much more deeply into our conception of science than is usually realized. On the outside there is the general political issue of identifying socially desirable goals and putting science to work on them. The resolution of this issue then becomes one of those familiar melodramas in the theatre of social power – the conflict of partisan classes, of ideologies and values, of economic theories and commercial realities, of democratic against hierarchical government, and so on. At a more mundane level it involves research budgets, management structures, advisory committees and the other paraphernalia of the art of administration.

The very idea of a 'Science Policy' (§5.7) to achieve large and distant social goals tends to divert our attention from the high proportion of scientific and technological effort that is being devoted to industrial development for purely commercial ends. That is to say, at least a half of what I have loosely called the R & D *system* consists of scientists and technologists working in separate research laboratories on behalf of corporations whose main purpose is to produce things that people will want to buy and use. One of the most controversial STS issues is whether the technological innovation that arises through commercial competition of this kind is as beneficial to society as a whole as many people think, and whether it should be subjected to much sharper *technology assessment* of both benefits *and* risks over a longer period. Where the proposed 'benefits' are themselves highly questionable – as they must be in the assessment of any military project, for example – the social role of science itself comes under severe scrutiny.

These issues are undoubtedly of the greatest importance. It matters to all of us whether scientific resources are heavily committed to military projects, or whether they are available for more peaceful industrial development. The difficult policy choice between different energy systems will have its counterpart in the relative emphasis to be placed on research on nuclear and other power sources. Failure, until recently, to mobilize large-scale research on environmental pollution can be interpreted as a political tendency not to interfere with the operations of industrial corporations. The subordination of much scientific and technological effort to purely commercial considerations, often to the detriment of human welfare, is a cause for grave concern.

The R & D system is a significant component of the national economy and polity. The way in which it is financed and managed, and the objectives towards which its major units are deployed, are issues that come near to the centres of power in our society. But they cannot be resolved solely in financial, managerial or 'public interest' terms. As the range of policy extends beyond the immediate technological horizon, the more fundamental dimensions of science become involved.

The question whether a particular 'end' is likely to be achieved in practice by a particular 'means' can only be answered in the light of the knowledge that is currently available. Consider, for example, the immense efforts now being exerted to construct sources of fusion power by means of a magnetically-confined plasma. This project is quite unprecedented in human history. Very considerable resources are being devoted to it, over a long span of years, in the hope of arriving at a highly practical objective. And yet that objective – a commercially viable source of energy – will not be reached for at least another generation, and may yet turn out to be an extravagant mirage. The economic and political analysis of such a project makes a splendid subject for STS studies. But the real question, at the heart of such analysis, is whether it is technically feasible, in the light of what is currently known. A great deal of the information that is needed to make such a judgement is essentially private and personal and only to be found amongst little groups of experts working in various national research laboratories around the world. But a major part of it is already stored for reference in the archives of public science, often stretching a long way into the past. The way to fusion power is not solely through the application of technical skills in electrical engineering and vacuum plumbing: it springs from Maxwell's equations of electromagnetism, and Einstein's mass–energy relation, it climbs up many dizzy academic staircases of nuclear and atomic theory, magnetohydrodynamics, laser optics, etc. The long-term policy issue must be judged in terms of the way in which this knowledge was arrived at and the extent to which it can be relied on to make predictions a long way into the future. Have the theoretical alternatives to the Tokamak configuration been adequately explored; is there yet a satisfactory analysis of the different modes of plasma instability; can we really be sure that radiation damage to the confining structure will not make all such designs hopelessly uneconomic?

The instrumental conception of science must take the 'philosophical' or 'cognitive' dimension (§4.5) into account. It must be concerned about the procedures by which scientific information is acquired, communicated, tested, reformulated, and reconfirmed, until it can no longer be seriously doubted. In every conceivable application of science, there has to be included a large safety factor, or margin of uncertainty, just for the errors that will undoubtedly be found in supposedly well-established knowledge.

The personal and sociological dimensions must also be included in the analysis. The scientific books and journals in the academic libraries cannot speak for themselves: their contents have only the value of what can be found in them by skilled scientists (§4.3). No question concerning the proper use of science can be decided without recourse to the specialized judgement of scientific experts (§6.3). Science is not so much a *passive* instrument of thought, as an *active* human agency, entrusted with responsibility for certain tasks. It is of the essence of research that it should be carried out by independent-minded, self-winding people, working in quasi-autonomous institutions. The interpretation that these experts put on scientific knowledge is inevitably coloured to some extent by their more personal concerns as research workers, their experience in their careers, their relations with their scientific colleagues, and other factors that cannot be ascribed simply to general political and economic forces.

It often turns out that potentially applicable lines of research are not pursued, or science is not put to optimal use, for reasons to do mainly with the peculiar structure of the scientific community (§4.4). The limited vision of some leading 'authority', reluctance to abandon a 'well-established' paradigm, rivalry between different 'schools' of thought, or heavy intellectual investment in a 'methodology' may seriously distort the research process. Support for basic and strategic research is often determined by peer review procedures that give much more weight to 'internal' factors, such as the state of knowledge in a particular discipline, than to 'external' considerations such as the conceivable applicability of what might be discovered.

In the long run, small decisions that are mainly determined by forces in the sociological dimension of science may have a greater influence on the course of human affairs than highly visible 'big' decisions such as whether to proceed with an expensive prototype of power reactor or aircraft. If we believe that science should be put to the best possible use, then this is a general goal that must itself be part of the ideology of research. The way in which science gets to be applied depends upon the climate of opinion on such matters amongst scientists themselves. It also depends upon the perceptions of the policy-makers and their advisers concerning the nature of science. If science is thought to be entirely rational, technical, divisible into specialized skills and applicable in principle to all aspects of human affairs, then emphasis will be on technocratic procedures to find a 'technological fix' for every problem. The scientism of conventional science education (chapter 3) generates its own style of technology, with its own characteristic orientations, deficiencies, and misuses. The R & D system is remarkably self contained and closed in on itself; what scientists and technologists believe about the dynamics of putting science to use is part of that dynamics; the educational system is one of the few input terminals to this peculiar black box.

6.3 The scientist as expert

The R & D system is the repository of a vast amount of potentially useful information. If you want to know the facts, ask a scientist! In our modern technological civilization, scientists are the supreme *experts*, whose specialist advice, opinion or wisdom must be sought on innumerable practical questions that go far beyond the planning of scientific research.

This social role is played in public, for example, by a professor of engineering appearing as an expert witness in a legal case concerning the performance of a bridge or aircraft. Engineers, physicists, and radiobiologists give important testimony at public enquiries on the siting of nuclear power plants. After any technical disaster, such as an explosion in a chemical plant, the views of scientific experts are widely publicized in the news media. And every government agency or industrial corporation has its advisory committees, consultants, and 'in-house' experts whose views are called upon (if not always followed) on a variety of issues concerning current operations and future programmes.

Since the R & D system employs a horde of people with every conceivable specialized qualification, there should not be any great difficulty in getting the advice that is needed. According to the positivist stereotype of science (§3.3), the practical world poses 'problems', to which the scientist gives rational 'model answers'. These answers have the peculiar virtue that they are 'objective', and hence free of the 'values' that prejudice the advice given by politicians, lawyers, businessmen, clergymen, and other interested parties.

Nevertheless, the deficiencies and discrepancies in the expert opinions of scientists on practical matters is one of the characteristic themes in the social relations of science. The technocratic ideal (§3.7) of rational, objective, unprejudiced advice seems never to be realized in practice. In public and in private, scientific advisers turn out to have many of the defects of other experts, together with a few special vices of their own. The 'QSE' (Qualified Scientists and Engineers) category is much more diverse, and its members much worse qualified to rule society, than is allowed for in the scientistic ideology. To understand this situation – and hence to be in a position to make the best use of the scientific and technological expertise which is now absolutely essential to most sophisticated undertakings – one must look more deeply into the R & D system.

It is obvious, for example, that a qualification based solely upon formal education and examination means practically nothing. As in all skilled crafts and professions, there is an immense range of competence not only between the raw recruit and the veteran, but also amongst fully experienced practitioners. Scientific knowledge is so far from being 'objective', that it is almost invariably sought from a leading 'authority' on the subject rather than from any person with a Ph.D. The social role of the scientific expert is confined to a very small élite of notables; there is very little call upon the knowledge or skill of those who have not achieved a high scientific reputation.

In practical professions, such as clinical medicine or design engineering, the worldly wisdom that the best advice is to be bought at the highest price from the top person in the business is a rational policy. A consultant physician or engineer usually owes his reputation to the success with which he has given advice on similar issues in the past. He may not be an original genius, but can be relied upon to have a good understanding of the practical circumstances in which his opinion is to be applied, and the likely outcome of various policies in such a situation.

But when it comes to more speculative issues, leading further into the future than can be extrapolated from current practice, the only available experts are those who have specialized in more academic, more hypothetical knowledge that has not yet been tested by use. For advice on the prospects for genetic engineering, one must go to a highbrow molecular biologist; for a proposal on geothermal heat supply, one would probably have to consult a professor of geology with no experience of power plant design. It is only through such people that there is access to the large body of systematic, ordered, public knowledge in the scientific archives – knowledge that generalizes and transcends the particulars of the technological practice of the day.

At the academic core of the R & D system, one encounters a type of professional authority quite different from the expertise of the consultant practitioner. Nobel prizes and other badges of scientific notability are awarded for talent at making discoveries, not for breadth and soundness of knowledge, nor for the shrewd application of what is known. The scientific élite is not selected for competence in assessing the credibility of technical information, nor for giving practical advice. As can be seen from study of the internal sociology of science (§4.4), and the psychology of scientific creativity, the personal traits of a leading scientist and the norms of behaviour to which he or she has been socialized may be quite inappropriate to the role of an expert adviser in practical affairs. Within the scientific élite (as in any other group of intelligent, strongly motivated people) there are to be found quite a number of individuals with the abilities for this role, but outstanding success in research is no guarantee of political wisdom, practical good sense, moral probity – or even sound judgement on purely scientific matters. It is only the naive and simplistic image of 'valid' science presented by conventional science education that would suggest otherwise.

Most ordinary people never have any opportunity to choose their own scientific experts. This choice is made for us by the political authorities who appoint science advisers, members of commissions of enquiry and other suchlike pundits. But the credibility or practical reliability of specialist scientific and technological advice concerns us all, both as citizens and in our private lives.

Here, above all, it is vital to understand the research process from its technological manifestations right down into the academic core. One must be clear, for example, that there is no worthwhile alternative to well-informed scientific opinion where it is properly applicable: there is no independent source of authority, no procedures other than the methods of science itself, by which more reliable knowledge can be arrived at and validated in defiance of the experts. For all its weaknesses, genuine scientific advice cannot be ignored.

On the other hand, most scientific experts have only a very limited range of reliable knowledge. A typical scientific speciality may cover no more than a few per cent (so to speak) of a conventional 'discipline'. Thus, even a distinguished expert on the theory of the atomic nucleus with the title of 'professor of physics', would not usually be well enough informed to offer practical guidance on the physical properties of semiconductors or the behaviour of a plasma in a magnetic field. Outside physics – in genetics, for example, or pharmacology – his opinions would scarcely be of more value than those of any other scientifically educated person. The real strength of science lies in the accuracy of precise, well-tested details, not in some peculiar wisdom accessible only to scientists.

Those who turn to science for reliable knowledge are often baffled by a plethora of conflicting opinions. Some scientists say this and some say that. In place of objectivity and certainty, the enquirer encounters controversy and doubt. Sometimes this conflict of opinion can be traced to factors outside of the primary knowledge system, such as political forces or religious prejudice; but usually it is just a reflection of the characteristic state of any active field of research.

Even at its academic core, the R & D system is not founded on absolute certainty. The expert authorities may be almost unanimous about a certain number of basic, well-established principles, but these can seldom be invoked to give definite answers to the questions that arise around any practical issue. A conjectural technological innovation whose assessment is being attempted is usually postulated out of the findings of contemporary basic or strategic research, which is itself in a state of intellectual anarchy and rapid change. How can any scientist, for example, advise on the future of fusion power without reference to current research on plasma physics, where there is such richness of inconsistent data and debatable theory? In many cases, the questions to be answered by the scientific experts have never been thought of before, and lie far outside the scope of the current consensus – witness the attempts to predict the effects of supersonic aircraft on the ozone layer in the upper atmosphere, or the permeability of deep crustal rocks to radioactive wastes.

STS education can never expect to give the correct answers to all questions of this kind. Indeed, we can be quite sure that the controversial issues of today, such as nuclear power and environmental pollution, will have been entirely superseded by others within a few years. Scientific opinion cannot be summarized and learnt for future use like a list of irregular verbs in a Latin grammar. But by showing how science really works, by giving some idea of the way in which the reliability of scientific knowledge changes during the process of validation, by talking about how scientists are trained, how they do research, and how they arrive at positions of eminence, we may hope to protect society from some of the elementary pitfalls in the use of scientific expertise in practical affairs.

6.4 Social responsibility in science

Public concern about the social role of science thrusts a considerable amount of personal responsibility on to the professional scientist. This is not a burden that many people are eager to take up when they enter this profession, even though it may come to weigh quite heavily in later life. The ethical dimension of research is entirely ignored in conventional science education; yet this is one of the most important issues for the scientist in relation to society.

The traditional ideology of science (§3.8) divorces the *scientist* from all such responsibility. His (it would be an entirely male profession in this ideology!) involvement in issues outside of research is solely as an aloof, disinterested expert, responding objectively to questions of fact. Otherwise, he is the epitome of a free intelligence, a natural philosopher, a *savant*, quite detached from the demands of society, following only the light of his own curiosity, communing with kindred spirits, and eventually making his learning available to a grateful mankind. It is recognized, of course, that scientific knowledge gets widely used, mainly for good, but sometimes (alas!) for ill. But the responsibility for the disadvantageous consequences of a scientific discovery is held not to rest in any way with the scientist who made it – should the parents of a new born infant be held responsible for the criminal acts that it might commit in adult life? – but with whoever applies this knowledge. According to this ideology, the body of scientific knowledge collected in the public archives is as ethically 'neutral' as a mountain or a lake, and those who produce this knowledge are entirely blameless for any evil use to which it might be put. Indeed, since science has evidently given mankind untold benefits, there is moral credit to be gained from taking part in research – to the extent that the pursuit of knowledge is held to be absolutely virtuous on its own account.

How, then, people ask, can one account for the very large amount of scientific research that is being carried out quite deliberately for inhuman or morally disreputable ends. In the conventional scientific ideology, the blame for this is made to fall on the 'applied scientist', the *technologist*, who draws on basic science for its technical applications. But this blame is easily shifted on to less personal shoulders. Almost all technologists are employed as professional experts by powerful corporations, over whose policies they have practically no control. The R & D system (§5.5) is an instrument of the power centres of society – the government, capitalist industry, the military, etc.; the prime duty of the technologist is to serve this system competently and loyally by producing the research results or carrying out the technological developments required by his (of course!) masters. He may at some stage have had to consider whether he should seek employment in this or that corporation, and thus accept its ends as his own. But from that point on, he can feel free from any personal responsibility in the orientation and later consequences of his work. If this means that he must take part in the development of napalm, then the technocratic view is that he is no more to be blamed than the salesman who hawks the stuff round the nations, or even the poor devil who eventually has to use it on the battlefield.

These arguments obviously have their own weaknesses. As any student of ethics could easily demonstrate, the servant of an evil master is not absolved from personal responsibility just because he obeyed orders, and he who knowingly provides the weapon is also guilty of the crime. It is striking, moreover, that the premises of these pleas are entirely contradictory of one another and effectively cancel each other out. The dichotomy between the 'scientist' who responds only to his own needs in the intellectual realm and the 'technologist' who responds only to the demands of others in the practical sphere, is totally false.

This illusion is maintained by treating academic research as something quite distinct from technological development. The fact is, however (§5.4), that the boundary between 'science' and 'technology' within the R & D system has practically disappeared. Modern academics are much more deeply implicated in the applications of their discoveries than they ever were in the past. By accepting the material facilities that they think they need for research, they cannot avoid some degree of subservience to the organizations from which these resources flow. Government funds that are supplied according to social, economic, industrial or military policies (§5.7) carry the ultimate goals of those policies into the university laboratory, seminar room and study. Such goals may be perfectly proper in themselves: in time of war, for example, patriotism is usually regarded as the ultimate virtue. But the responsibility for supporting them cannot be shrugged off on the grounds that the product of research is just 'neutral' knowledge whose use cannot in any way be foreseen.

On the other side of this vanishing line of demarcation, the scientific technologist has more interest in fundamental research than ever before. Not only should he or she respond in full to the demands of humanity and social justice in the course of his or her employment; there should also develop some concern for the conditions in which the 'disinterested search for truth' can in fact continue. It may come to be seen that independence of mind and thought, freedom of criticism and enquiry, and the other characteristic ethical commitments of academic science, are the foundations of the knowledge system on which technological R & D is really based (§5.4). Concern about wider, longer-term, social needs may thus be transformed into a personal feeling of responsibility for knowledge and truth as values in themselves, in conflict with a more expedient loyalty to a company or a country.

This is not the place to argue such matters out in detail. Ethical dilemmas can never be properly resolved by reference to blanket principles, valid for all persons on all occasions. The answer that any particular individual gives to the voice of conscience depends upon many different personal factors – patterns of upbringing, cultural and religious principles, psychological and emotional make-up, etc. In a free society, freedom and diversity of ethical response is the fundamental liberty.

But the ultimate irresponsibility is to refuse to face up to these issues at all, or to fall back, without further reflection, on one or other of the extreme positions, as 'scientist' or as 'technologist'. Now that the whole question of social responsibility in science has been brought out into the open, it cannot be hastily packed away in case it caused a scandal! The full significance of these ethical dilemmas and conflicts of personal loyalty cannot be appreciated, nor can the typical situations in which they might arise be rehearsed, without reference to the subtle internal structure of the R & D system, and the complexity of the social roles associated with research as a profession. For the would-be academic scientist, for the would-be technologist, as for the most innocent lay person, there is great need for a reliable representation of the intellectual, institutional, professional and emotional background against which such modern dramas of conscience as the personal ordeals of Robert Oppenheimer and André Sakharov are actually being played.

6.5 Science as a part of culture

As a final example of the sort of theme that is frequently discussed in STS education, let us consider the place of science in modern life. This is, inevitably, a very vague and ill-defined topic, that easily runs away into wild irrelevancies; but for many people, young and old, it is of great interest and concern. It cannot be spanned by a single direct question, but forces us to look deeply into the whole research process, peeling off layer after layer until we reach the central cognitive core.

From the outside, the issue is often posed as a critique of the technological style of life (§5.3). Attention is focussed on the sophisticated products of scientific technology – microcomputers, supersonic aircraft, nuclear weapons, 'automated' medicine, etc. – and questions are asked whether the further development of such devices is favourable to human happiness, social justice, or other values of life. Since this trend in our civilization is obviously dependent on the R & D system, it can be immediately attributed to the influence of science – which is then to be praised or blamed accordingly.

To discuss our theme at this level, it is essential to understand how the R & D system works, how it is directed and financed, how far it originates technological innovation, and how knowledge gained by research is exploited in practice. Many of these features of the R & D system can only be discussed by reference to the more general political, economic, and social aspects of modern life, which cannot be dealt with adequately within STS education. But the science/technology complex is certainly not a passive factor in economic activity, and would play a vital part in any large change in our civilization, such as the transition to a 'steady-state' culture based on ecologically renewable resources. Is the radicalism and competitiveness of research activity compatible with a less dynamic style of life? Does the scientific mode of technical change inevitably stress extravagant, overblown, inhuman ways of living? Is new technology 'pushed' into society by science, or is it 'pulled' out by public demand? Can we resist the 'technological imperative' (§3.5) that 'anything that can be done will be done – must be done'? These are typical questions that arise when we consider the place of science-based technology in contemporary culture.

But deeper questions then begin to be asked. For the majority of people now on Earth, the quality of life is not much to be proud of. There are dreadful material and social deficiencies – hunger, disease, tyranny, environmental degradation, unemployment, exhaustion of resources, and so on – from which, one hopes, billions of people could be freed. Surely science can show its value by being directed towards the relief of such conditions.

In our modern culture, science is regarded as the means by which all manner of problems are to be solved (§5.1). The characteristic response is to undertake research, out of which will come, in due course, a programme of action that will cure the disease, improve crop yields, find alternative resources, halt inflation, or even stop wars. But this approach to the human condition, implicit in much of science policy, can be seriously questioned. What may come out of it, for example, may be no more than a 'technological fix' which ameliorates the condition without eliminating the root causes. Is science the most effective means – perhaps the only means – of making a better world? Does it deserve this privileged position in our society?

To deal with this sort of question, we must consider the potentialities of a 'scientific' approach to a much wider range of situations than in the past. Western society has tended to apply science only to physical and biological problems where there is a good prospect of material gain. Can the same methods be applied successfully to problems where socio-economic factors – such as food preferences and systems of land tenure – are of prime importance? There has long been a movement for 'scientific rationality' in our political and social arrangements to match our technological achievements. But this movement has always come up against other systems of belief, with ethical or religious foundations. In any assessment of the place of science in modern life, one cannot avoid the question whether all such alternative sources of belief should be dismissed as essentially irrational (§3.4), or whether, on the other hand, many of our cherished goals are just as likely to be reached by the exercise of practical wisdom, political skill, moral rectitude, and other such traditional virtues as by the more scientific process of defining a series of problems and seeking their solutions.

And here, of course, we trespass into regions of political and religious controversy that are normally forbidden to teachers and students of science. The proposal that the human condition should be ameliorated by the application of science is not as neutral as it is made out to be. It impinges on the structure of society itself – who gets what out of it, who cuts the cake, who decides priorities and procedures. You cannot talk about the applications of science without taking up attitudes in the political sphere – although the political scientism and technocracy often found amongst the mandarins of high science has both its right-wing and left-wing manifestations. Many science teachers shy away from such topics, fearing that they may affect the 'validity' of their subjects. But these controversial issues, and the conflicts between their supporters, are amongst the most important features of the social context of science, and can no more be excluded from STS education than from disciplines such as economics and political theory where they are discussed more centrally. They should be treated evenhandedly; they cannot be shut out of the classroom or seminar just because they must not be taught in the dogmatic style of established science itself!

Nevertheless, it is important not to politicize every question about the place of science in our culture. One must also consider the nature of scientific expertise, the way in which research programmes are formulated, and the fundamental capabilities of the scientists who are expected to carry out this research. It may be, for example, that the training necessary to make any 'valid' contribution to knowledge is so specialized (chapter 2) that it makes people almost incapable of appreciating the larger, more subtle issues that are really at stake in the world of affairs. There may simply be no way of melding the diverse insights provided by various specialized disciplines into a single description of the situation (§§3.2, 4.5), from which might be evident the policies needed to improve it. A study of the research process as a means of solving problems might show that socio-economic issues are always so complex and convoluted that they are beyond the scope of rational analysis and prediction. It might be argued, for example (§3.9), that the scientific ideal of designing and running a large corporation or national economy according to the same general principles as one would design and run a jumbo jet or a petroleum refinery is an absolutely inappropriate metaphor, and that intuitive practices of personal leadership or collective decision-making are not to be despised.

In the end, it is impossible to avoid the question whether there are realms of existence and action that are, of their very nature, not amenable to the scientific approach. What are the prospects of ever constructing a science of human behaviour that would be in any way comparable with our present sciences of matter and of life? Quite apart from the practical question whether such a science would be of real help in solving complex psychological and social problems, we must seriously consider whether the scientific approach is valid in principle when applied to even the very simplest phenomena involving conscious beings. In other words, how wide should be the range allotted to scientific modes of description and explanation within the intellectual culture of our era? Is there, for example, a permanent line of demarcation, or at least a buffer zone, between the natural sciences and the humanities, corresponding to a distinction between 'facts' and 'values', or between 'objective' and 'subjective' points of view?

To proceed with such an investigation, we must study in some detail the way in which basic scientific knowledge is actually obtained. Attention must be concentrated on the inner academic core of the R & D system, where the goal of research is to collect reliable knowledge about the external world, without direct reference to possible use. In this process, the part played by the individual scientist, as observer, communicator and rational critic cannot be ignored (§§4.3, 4.6). It may be argued, for example, that all social science implies some interaction between the observer and the system being studied, so that the canons of scientific objectivity cannot be obeyed – to which the response might be that all scientific observation suffers to some extent from this defect, which cannot therefore be regarded as an objection in principle against a science of behaviour.

At first sight, such matters seem very academic and irrelevant to our social concerns. And yet it turns out, so very often, that what is really at stake is the status of science itself. The cultural role of science depends upon whether it can substantiate the claim to be more reliable, more objective, more 'true' than any other form of knowledge. Everyone concerned with science in practice must respond somehow or other to this epistemological challenge. This response cannot be entirely negative. We may decide to reject a 'scientific' technology because it does not provide us with what we really want; we may reject a 'scientific' solution to a problem because it does not seem to allow for various intangible human factors; we may be sceptical of a 'scientific' explanation of a social event because it does not take account of the self-consciousness of the persons involved; but we take our lives in our hands if we reject the calculable consequences of a 'scientific' law such as Newton's equations of motion, or refuse 'scientific' treatment with antibiotics for pneumonia.

There can be no serious doubt that some of our scientific knowledge is as correct as any knowledge could be. Does this mean that it is absolutely true? To answer such a fundamental philosophical question, it is not sufficient to look only at the contents of the scientific archives (§4.5). This study demands a scrupulous regard for the realities of the research process – the psychology of invention, the neurophysiology of observation, the techniques of experimentation, the logic of theorizing, the media of communication and criticism, and the norms of the scientific community where these activities take place.

It would certainly be quite improper to propose a specific answer to this enquiry. Indeed, what the student will mainly learn from pursuing it to this depth is the wide range of credibility – and practical reliability – of what passes for scientific knowledge. It is just as important to realize the uncertainty (§6.3) of a great deal of present-day science as it is to have confidence in at least some of its major discoveries.

There is also much to be learnt about the nature of scientific theories, and the way in which they seem to map out for us an ultimate reality. In other words, the place of science in modern life (§1.5) may be that it provides the 'world picture' around which our whole culture is articulated. But we may legitimately ask whether this picture is coherent and complete (§3.2) – or whether, as many people suspect, it is only a fragmentary view of the way things are.

It may be, indeed, that even if a coherent view of the external world could be given by modern science, its claim to absolute truth might not be valid. It can be argued that quite a different view of things could be regarded as equally convincing in a different social order – if not in any past or present human society, then perhaps in the distant future, or on some distant planet of another star.

Philosophical relativism is almost impossible to refute in principle, and can be given some empirical support by reference to the influence of social relations and material conditions on the direction and outcome of scientific research. It thus turns out that a theme that has tunnelled deeper and deeper into the logical and psychological foundations of the R & D system, until it seems to be concerned mainly with the abstruse practices of academia, raises subtle and sensitive questions concerning the broadest aspects of our subject – the relationship between technology and social structure, and between ideology and social power.

These familiar issues – science and human need, the value of scientific expertise, social responsibility in science, science and ideology, etc. – illustrate my basic theme that the R & D system should not be treated as a 'black box', without reference to its constituent elements and their complex interactions (Fig. 6). At every level, whatever the issue we raise, the response of the system as a whole must be taken into account.

But, as I hope this chapter has shown, there is a coherence in this model that knits together a number of threads connected with the social role of science to an extent that is not always apparent in more partial and limited schemes. This is also a framework within which the personal experience and real beliefs of working scientists and science teachers (§3.9) can be explored seriously behind the traditional scientistic facade. In this frame of mind, STS education can be seen as a natural adjunct to conventional science teaching, and not as antipathetic to science as such.

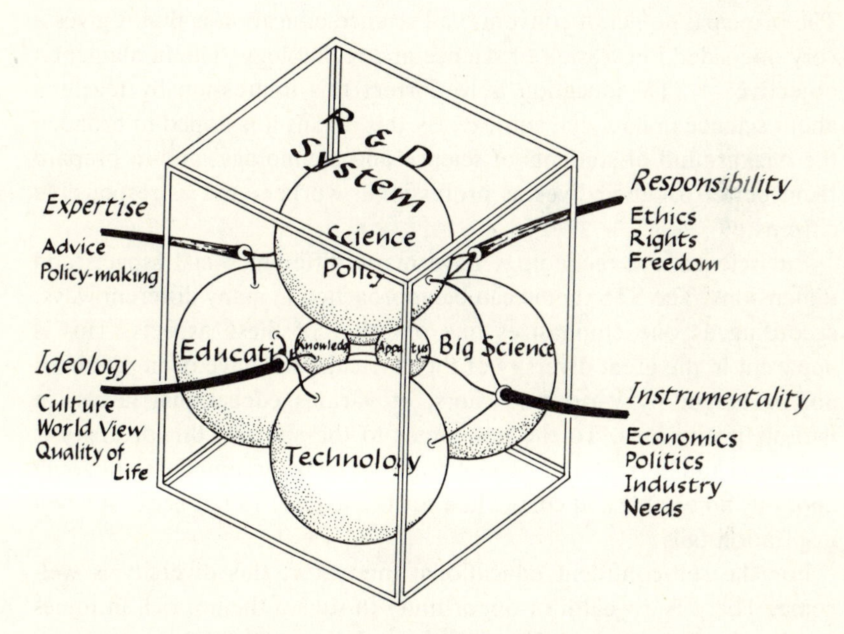

Fig. 6. Science in its social context

7

Approaches to STS education

7.1 A diversity of ways

The principal defect of conventional science education is that it gives a very one-sided impression of science and technology. The fundamental objective of STS education is to correct this impression by teaching about science in its social context. By this means it is hoped to broaden the background of students of science and technology, and to prepare them better for their lives as professional workers and as responsible citizens.

But science and technology have many different social aspects and dimensions. The STS theme can be approached in many different ways, according as one emphasizes one or other of these aspects. This is apparent in the great diversity of the curricula that have been proposed and/or taught by various teachers, at various educational levels, in various institutions. To the newcomer to the subject, this diversity is confusing and intimidating. There seems to be no accepted way of going about it, no established curriculum for the teacher to fall back on when inspiration fails.

For the self-confident educational innovator, this diversity is welcome. There is a wealth of opportunity in such a theme, rich in topics and in educational styles. The STS theme has something for everybody, from pedantry to fantasy, from austere abstraction to opulent reality, from aloof analysis to committed concern. It draws on philosophical, sociological, psychological and historical disciplines, as well as all the sciences and technologies. It can be concentrated on cognitive questions, on political issues, on technical capabilities, or on moral judgements. Unlike the more sophisticated scientific disciplines, it is not an intellectual pyramid (§2.1) whose base must be climbed laboriously before one can reach an interesting point of vantage, so that the curriculum is not narrowly constrained by the knowledge that is regarded as 'prerequisite' to each successive stage.

It is an integral part of the philosophy of STS education that it should *not* be tamed and domesticated into a conventional academic curriculum. Much of its value to students – and to teachers – lies in the openness, divergence, uncertainty, and spontaneity that it can bring into the classroom and laboratory. The legitimate topics and modes of teaching cover so wide a range that a great deal of responsibility must be left to the teacher to choose material and methods suited to the intellectual, social and psychological maturity of his or her students, and to their intended careers.

Nevertheless, this pedagogic freedom can easily lapse into educational anarchy. As has already been pointed out (§4.1), the STS movement gives the impression of being more united in its opposition to the way in which science is conventionally taught and practised than in any positive view of its chosen subject. Anyone who claims to be teaching about science and society should have in mind at least some more or less coherent general idea of what they think they are talking about, and how their own particular approach is related to various other aspects of the subject. Without some such global view, the whole STS theme fragments into a disconnected assembly of specialized academic topics, pedagogic exercises and ideological doctrines.

The 'model' of the R & D system outlined in the last three chapters is the sort of theoretical framework that is needed to bring the subject together. As I have repeatedly emphasized, this particular scheme would not be accepted in detail by every science teacher or by every supporter of the STS education, but it does cover most significant aspects of the subject and represents their interrelations in a credible way. It would, however, be a serious mistake of educational strategy to expound such an abstract general model of science as an introduction to STS studies. Not only would it be improper to try to impose this interpretation in advance on the various particular topics and issues that need to be studied: any such effort would almost certainly fail to draw the interest of the students and to hold their attention. Purely as a matter of practical pedagogy, the STS movement cannot afford to discredit itself by pushing a highly academic and unteachable curriculum on to science education. The general theory of science, in its philosophical, psychological and social aspects, is a perfectly proper subject for study and research, but only at an advanced level (§8.9) after considerable acquaintance with various actual cases illustrating particular theoretical and practical issues.

We may agree, then, that there is a single entity – the R & D system, with its core of academic science – to be taught about. But where is one to begin, by what pedagogic strategy, to attack and penetrate to the interior of this 'black box'? Having chosen such a strategy, are we then in peril of overemphasizing one or another aspect of the STS theme, and thus fail once more to indicate its essential unity?

Perhaps the best that can be done to avoid this pitfall is simply to be well aware of it as a danger along any path we choose to follow. The diversity of topics and styles in STS education is a grave weakness only if it is a symptom of fundamental disunity and fragmentation of knowledge, opinion and objectives amongst those who teach it; otherwise it is a sign of intellectual vitality and educational sensibility in a new and growing subject.

In this chapter, therefore, we shall consider some of the very many different ways in which the study of science in a social context may be approached. To avoid a lot of tedious and essentially ephemeral detail, there will be no attempt to set out any particular curriculum as now taught, or as proposed, nor will there be any discussion of the successes and failures of specific programmes. This may sound a little abstract and academic; why not provide a critical catalogue of STS curricula, with useful tips for the beginner? But it is not a practical handbook that is needed – it is a 'philosophy', a 'rationale' by the light of which any particular proposal can be analysed. Just as there is a need for a more conscious awareness of the general nature of science itself, so there is a need to bring to the surface the goals and capabilities, the advantages and disadvantages, of these various possible approaches to the subject. This analysis can be no more than schematic, not because a much more detailed and particularized analysis would not be useful, but because this is a territory that has not previously been explored and mapped overall.

It is quite inevitable that any analysis of educational goals and methods will be strongly biassed by the personal values of the analyst. Some theorem or another of the sociology of knowledge informs us that this bias can never be fully compensated, however sincerely the effort is made. Nevertheless, in what follows some allowance should be made for my own prejudice against using the STS theme as a vehicle for dogmatic indoctrination (§4.2). The fundamental political and ethical pluralism of the open society is taken for granted.

In my opinion STS education should not be exclusively for, or against, science; for, or against, technology; for, or against, industry; for, or against, the government; for, or against, capitalism; for, or against, socialism – not even, in the current political double talk, for, or against, 'people', or for, or against, 'peace'. On the contrary, it should be seen as the means by which diversity of opinion, tolerance of controversy, imperfection of decision, and unpredictability of the outcome of action can be illustrated to the student through a medium that is often supposed to be immune from such maladies. The main thrust of the STS movement is to oppose scientism and technocracy: it must reject any similar narrow formula that pretends to know all the answers to all the problems of our times.

Another, more subtle trap for the earnest teacher is to suppose that he can teach his pupils *wisdom*. With a wider view of life, longer experience, greater social maturity, he or she comes to believe that, like a well-intentioned parent, one can guide young people through the rocks and rapids of life. Like most parents, one is bound to fail. Wisdom may be acquired to some extent by imitation of a good example, but it cannot be *taught*.

In the preamble to every proposed STS curriculum, numerous highly laudable educational aims and objectives are set down. But mostly these are the 'aims and objectives' of education as a whole (§1.1), and are already engraved on the heart of every teacher. One of the virtues of STS education is that it offers occasions to practise some of the 'humanistic' skills that are largely excluded from the hard sciences. This is a very good general reason for introducing the subject into science education, but it should not be exaggerated into the sole pedagogic principle around which every STS curriculum is to be designed.

We can provide our pupils with valuable information; we can give them various opportunities to exercise and strengthen specific skills such as arguing, talking, writing and reading; we can make them more knowledgeable and more expert; we can even, if we are sympathetic and understanding, generate around them an atmosphere that fosters latent talents and opens doors and windows into wider worlds of thought and feeling; but we cannot, as mere teachers, make them anywhere near as wise (or, thank God, as foolish) as ourselves.

7.2 Making valid science relevant

The most natural approach to society from science is through its applications. That is to say, the conventional curriculum is extended from teaching valid science towards explaining its social relevance. Lectures on the theory of electromagnetism give their attention to the electronics industry and telecommunications; nuclear physics leads on to a discussion of nuclear weapons; chemistry courses deal with fuel cells or the effects of fluorocarbons on the ozone layer in the upper atmosphere; almost every topic in biology is shown to be relevant to medical, agricultural and environmental problems; even mathematics demonstrates its social applications in economics and ecology.

This approach comes very easily for the conventional science teacher or research worker. It also appeals to the interests and inclinations of most science students. There is the immense satisfaction of seeing theoretical knowledge turned to surprising use. It turns out that there is as much intellectual delight in a jet engine as in a law of thermodynamics. The intrinsic challenge of solving a practical problem, such as biological nitrogen fixation or the safe disposal of radioactive wastes, is discovered to be the equal of any fundamental research problem in molecular biology or high energy physics. It is only necessary to push on a little further, beyond purely technical questions to their economic and political correlates, and the STS theme is in full swing.

This pathway is broad and well-trodden. Indeed, as the class barriers between pure science and technology are dismantled, this has already become the characteristic pattern of much of science education (§1.3). The fact that basic science is largely being taught to students preparing for careers in technology is widely recognized. There is good practical justification within conventional science education for a shift away from the academic purism that was once in vogue.

But this new approach to the teaching of science may be little more than a shift of emphasis within the R & D system as a whole. As we have seen (§5.4), there is practically no line of demarcation nowadays between basic and applied research. Electromagnetic theory includes all the general principles of telecommunication engineering; the reactions that might occur in a fuel cell are typical of those treated in a course of electrochemistry; in academia itself, the 'biomedical sciences' are referred to without pedantic distinctions between those that are 'relevant' and those that are not.

In many instances, making valid science more relevant does not really penetrate to the social environment of the R & D system. For example, the calculations of the efficiency of internal combustion engines that might now find their way into a highbrow textbook of thermodynamics say nothing about the social effect of the automobile, or about energy resources, or even about research policy in the industries where this important item of physical theory finds its main use. It could be, indeed, a starting point for an exploration of any of these topics, but only if teacher and pupils collaborate in stepping right outside the technical frames of orthodox science and science-based technology.

Even then, it takes an act of high imagination to escape from the instrumental attitude (§6.2) which defines the social function of research as the solution of practical problems that arise in advanced technology, or as a breeding ground for novel concepts that will somehow get applied in advantageous ways. In the face of so much attention to what science *has* done, or *is* doing, there is little encouragement to think about what it *might* do, or *ought* to do. The fundamental orthodoxy of scientific and technological education converges on the established consensus, and has no place for wild speculation into unexplored territories of thought, or action.

The approach through relevance leads easily across the social map of science (chapter 5) from 'knowledge' and 'apparatus', through 'technology' and 'R & D', to the economics of industry and the technical aspects of science policy. But it misses altogether the personal and social dimensions of basic science and technological research. There seems no call to stop and look around at scientists themselves, as research workers (§4.3), as technical experts (§6.3), as members of invisible colleges (§4.4), or as personnel to be managed (§5.6). The whole question of the place of science in modern culture (§6.5) has already been answered by implication – in typical technocratic terms (§3.7). Because the STS theme is multiply-connected, these topics will no doubt be reached in due course; but they do not immediately impress themselves on the consciousness of the student along the way. In other words, this approach can scarcely avoid a heavily scientistic bias, where scientific 'validity' and technical 'capability' always remain dominant over needs, and values, and satisfactions, and ultimate goals.

7.3 The vocational approach

Science education is very largely a preparation for, or a part of, training for particular technological or technical careers (§§1.3, 1.4). Many students of science are on their way to becoming doctors, or nurses, or pharmacists or civil engineers or computer programmers or industrial managers. It makes good sense to include in their vocational training a substantial amount about the social aspects of their future jobs.

In the past, preparation for this aspect of professional life was largely neglected. The young technical worker was expected to pick it up in the course of a practical apprenticeship, as a hospital intern, say, or junior executive. But with the growth of new academic disciplines concerned with management, economics, social psychology, public administration, etc., there has been a trend towards including some such topics in the formal educational curriculum of technical trainees for many of the science-based professions. Established professional practitioners and industrial employers may be critical of the academic stance that is often taken in such courses, but the trend is encouraged as a way of preparing the student for life outside the ivory towers of school and college. There is a very practical, down-to-earth educational motive to 'tell students about industry', and to draw them into advanced technological professions, which now spreads right down into the secondary schools.

For many young people, indeed, courses of this kind may be very illuminating. They may come as the first significant encounter with a real world of obstinate materials, imperfect designs, and unregenerate people, outside the cosy comforts of home, school, friends, and theoretical concepts. These subjects are usually taught by social scientists who do not see things in the same light as science teachers, or by hard-headed people with practical experience who may take some pleasure in contrasting the harsh realities of commerce and industry with the idealized situations envisaged in conventional academic education.

This educational trend, therefore, is in the same general direction as the STS movement. It is very natural to approach the social context of science by a study of the vocational role of the scientist or technologist. In this way, for example, the many social issues that spring up around professional practice as an expert adviser or consultant (§6.3) can be introduced and analysed. The economic and administrative setting of high technology, where these parts are to be played, can also be sketched in. In other words, the vocational theme leads out of the technical dimensions of 'relevant' science and technology into the social and personal dimensions of the research system.

Further along this path, it opens up into some of the major ethical problems of science. The scientific expert must eventually face up to the great dilemmas of social responsibility, such as the choice between institutional loyalty and humane concern (§6.4). This becomes very obvious in medicine, where the practitioner is almost sure to encounter agonizing situations involving, say, the treatment of an incurable disease or the protection of a confidential medical record. The response to the STS movement within medical education is towards much more deliberate and open discussion of such life and death issues. There is little doubt that this sort of discussion could be extended to those students who are preparing for more academic research careers (§1.2), where there is less direct contact with living people; the vocational approach soon discloses moral issues that touch upon the personal rights and responsibilities of the scientist and upon the place of science in modern culture.

Nevertheless, this approach also has its limitations. Training for a specific career as a doctor, an architect, an engineer, or an industrial executive is necessarily, to a considerable degree, a preparation for a stereotyped social role. A curriculum whose technical content must be approved by a professional organization such as an engineering institution is not likely to stray very far towards a radical interpretation of the place of the engineer in society. Where professional employment (as in aeronautical engineering, for example) is almost entirely in large industrial corporations, the whole tenor of the education will be towards producing 'organization persons' who give little attention to the social consequences of the work they are paid to do. Or where, as in many countries, the physician is a self-employed entrepreneur with a personal 'practice', the social aspect of vocational training may be little more than indoctrination in the ideology of individualism and professional independence. The social awareness and moral self-criticism that many exponents of STS education would like to cultivate does not flow freely within this sort of curriculum.

The vocational approach may also reinforce the disciplinary specialization and professional exclusiveness of many career patterns in science and technology. By linking a whole curriculum with the practitioners of a particular technology or technique, it produces just the sort of narrow expertise that many people now regard with disfavour and regret. Instead of expanding the vision of the student to a view of the place of science, as a whole, in society, it is liable to contract it into a study of the part to be played by an individual member of a very particular social group, whose activities are not, in themselves, to be regarded as questionable.

7.4 Interdisciplinarity

The narrow specialties into which science so easily fragments (§2.4) have always been a target for reform. The history of science and of science education can be written as a chronicle of the appearance of new branches of learning and research, struggling to come to birth and to grow to maturity in the interstices of the established academic disciplines. In each scientific generation, the call for a new *inter*disciplinary approach to old problems, for new *multi*disciplinary curricula to meet future challenges, has been the rallying cry of scientific, technological and educational radicals.

As a programme for the reform of science education, the STS movement is often allied with – and sometimes confused with – this progressive tendency. The association is very natural. To achieve an interdisciplinary point of view, it is necessary to break down the barriers that enclose each traditional discipline within its intellectual paradigm, its technical methodology, or its problem area. Once these walls are breached many new topics, including the social aspects of the subject, can enter. If, for example, it is urged that physics and geology should open to one another to allow for the growth of the new interdisciplinary subject of geophysics, then this should facilitate other developments, such as the creation of new curricula on the physics of environmental protection, that touch directly on social concerns. The demand for a broader, multidisciplinary approach is evidently in sympathy with the STS critique of overspecialized scientific expertise and education.

It must be admitted, however, that this loosening up of traditional disciplinary demarcation lines is not necessarily favourable to the STS programme. As in the case of biochemistry, a new academic subject that lies across the borders of two older subjects may soon become as specialized and esoteric as its parents, eventually making similar claims to intellectual purity and technical virtuosity. The separation of scientific knowledge into distinct species, the differentiation of the scientific community into invisible colleges, which then hybridize, intermingle, and divide again into new groupings, is the characteristic evolutionary mode of science (§4.6). The attack on disciplinary constraints, in favour of a new interdisciplinary research programme, is often no more than a tactical move, whose strategic objective is to carve out yet another autonomous region on the academic map. This objective is, of course, entirely opposite to the goal of the STS movement which is to broaden and humanize science education.

The genuine spirit of interdisciplinary science is to be found in much humbler, less sophisticated educational programmes, under the heading of Combined Science, Integrated Science or General Science courses, where subjects are drawn from a wide range of the traditional fields of science and are not closely concentrated around a single problem or technique. Here there is fertile soil for the STS theme (§8.7).

Ideally, in such a course, the separate subjects of conventional science education can be related to larger themes which have significant social aspects. It is possible, in this way, to go somewhat further along the path towards social relevance than when one starts from a single 'valid' discipline. One can, for example, emphasize the powers of physics, chemistry and engineering, in combination, to deal with some of the problems of the energy crisis – and, by startling contrast, the impotence and irrelevance of all the hard sciences in the face of the social and ethical issues that then arise. One of the myths of scientism (§3.5) is that there is a 'science' (actual or potential) for dealing with every sort of problem. An exploration of the combined capabilities of the existing sciences soon shows the vast gaps and unknown regions of our scientific world view.

In fact, to make the most of the interdisciplinary approach it may be advisable to build the whole curriculum around some major practical theme such as 'energy', or 'the environment'. The real world knows nothing of our intellectual abstractions and categories; it exists fully in all aspects, and all dimensions. A natural object or a useful process can be studied in a multiplicity of ways, from the conventional hard sciences and technologies through the social sciences to the humanities. A curriculum founded on this principle would follow a famous educational tradition, exemplified by 'Greats' at Oxford, where classical culture is studied in all its aspects, linguistic, literary, philosophical, political, historical, etc. The concept of a 'Science Greats', a genuinely transdisciplinary general education, is strong as an ideal in the educational philosophy of the STS movement (§8.8).

And yet, it must be admitted, interdisciplinarity has its limitations as a dominant principle for STS studies. We may observe these in certain academic subjects, such as Geography, which already have many of the features of integrated science. On the one side there are topics such as climate and geomorphology, drawn from the natural sciences; on the other side, economic geography and town planning are drawn from the social sciences and humanities. A thorough education in geography inevitably crosses and recrosses the boundaries from science, through technology, into society without academic contrivance. Indeed, it is comic to observe that interdisciplinary subjects such as geography, or agriculture, or nutrition are forever being reinvented by naive science teachers, trying to introduce a note of 'relevance' into their courses. As Eric Ashby pointed out in *Technology and the Academics*, a thorough course of brewing technology has all the essential ingredients of a general transdisciplinary education.

But when we look in detail at the way such subjects are taught, we find that they usually take the traditional disciplines – scientific or humanistic – as their model, and are prone to the same deficiencies of overspecialization, academicism, and the exaltation of professional expertise. Like other disciplines, they do not reflect on their own development, and make no comment on research itself as a way of life and of art. The R & D system is taken for granted as an instrument of progress, as a means for solving practical problems, as an unquestionable source of benefit and intellectual authority. The path to a transdisciplinary point of view of many aspects of nature and of human affairs does not lead naturally into the heart of the STS theme.

Interdisciplinarity is a principle of great virtue in science education, but as an approach to the study of the social relations of science it is only a broader, more solid road from 'validity', through 'relevance' to economic, political, and other social issues. The place of science itself in the scheme of things does not come into it. In the next three sections, therefore, we consider how STS education may be approached through the direct study of science as a historical phenomenon, as an organized body of knowledge, and as a social activity or institution.

7.5 The historical approach

The Achilles heel of 'valid' science is its own history. Conventional science education is very vulnerable to the introduction of historical themes. Since history lends perspective to any view of science in its social context, radical STS reformers ally themselves with liberal-minded science teachers to advocate the enrichment of all science courses with historical material.

The orthodox academic scientist cannot resist this pressure, because research is pre-eminently an historical process. Its purpose is not to manufacture a permanent product, nor to reach a state of ultimate perfection: it is simply to generate change. The spirit of science is perpetual progress, continuous transformation. By definition, scientific knowledge accumulates and improves: it has always expanded: it has never suffered defeat. The triumphs of science do not celebrate peaceful eras of stability and abundance: they celebrate victorious discoveries, conquests of ignorance, and revolutionary epochs in which the image of the world is turned upside down.

Science is intensely proud of its own history. It misses no opportunity to remember the anniversaries of its heroes and to assimilate the achievements of present day notables to the traditions of their glorious predecessors. In cooler sociological language, communal recognition of past research contributions rewards significant communications to the archives, and reinforces the norms of the scientific community (§4.4). In more personal terms, old scientists love to reminisce, and get great satisfaction from the feeling that they are part of a noble enterprise stretching back into the past. Whether we look upon it as evolutionary or revolutionary, science would have no meaning for its practitioners if they were unable to perceive it in its historical dimension. Academic science necessarily supports, and desires, continual access to knowledge of its history.

The STS movement favours an historical approach for the same reason – to demonstrate science as the epitome of the spirit of change. The extraordinary self-transforming character of science is unique amongst social institutions, and diffuses outwards until it flavours all branches of material life and culture. The social, economic and political circumstances of our day can scarcely be comprehended except as consequences of the innovatory pressure of science and technology. In its social relations, science has deep and ancient roots which can only be traced by a return to traditional origins in academia. In other respects, however, there are fundamental changes of the social context and in the interrelations of the various components of the R & D system (chapter 5) – the centuries-long growth of the scientific community, the recent switch from academic to industrialized modes of research, the increasing scale of expenditures, of apparatus, and of science policy concerns, the deeper involvement of science in the arts of war and peace, and many other time-varying conditions.

The history of science is thus a favourite topic for inclusion in a reformed science curriculum. The only opponents in principle are tough-minded technocratic realists, who take as revealed truth all that they are at present teaching, or have personally discovered, or wildly imagine for the future, and who simply have no conception of the historical background of their current position. It takes an extreme and bigoted version of scientism not to see a good deal of wisdom in teaching about the past in science as a preparation for the future careers of scientists and technologists.

But what should be taught, and how should it be dealt with? The history of science is a subject of immense depth, diversity and richness, from which only an infinitesimal selection of facts and themes can be included in an elementary course. The historical approach to STS education is of unquestionable merit in principle, but not without severe defects and difficulties in practice.

In the first place, it is necessary to establish an adequate framework of primary factual knowledge. It is surprising how ignorant most students of science and technology can be of the barest outlines of the basic history of their own subjects. Ask them to assign a topic and an approximate date to famous names such as Newton ('mechanics, 1700', say) or Darwin ('evolution, 1850') and they will be quite stumped after the first 20 or so names, even in their own particular specialty. Various fabulous events or idiosyncratic details – the trial of Galileo, Einstein's aversion to socks – get handed on in the *obiter dicta* of science teachers, or in popular writings, but seldom with any accurate contextual information. School courses in economic and social history usually go to some length about the mechanical and chemical innovations associated with the Industrial Revolution, but have little to say about important developments in geology and biology that were taking place at the same time. It is idle – perhaps even mischievous – to offer an interpretation of the social role of science in history to students with such haphazard and fragmentary foundations of genuine knowledge.

Suppose that a course of instruction in these elements has been set up; where does it lead? Unfortunately, this sort of intellectual history is of infinite depth and complexity. For each great name, there are a dozen others worthy of mention. If we are to give an intelligible account of Newton's work, we must not only go back to Copernicus, Kepler and Galileo: we cannot neglect Hooke and Huygens, Descartes and Leibniz, Wallis the mathematician and Flamstead the astronomer. For each of these there is a life and works to be studied – and yet more figures are discovered, almost forgotten to posterity, but surprisingly perceptive and prolific for their times. This is the forest of detail into which the conventional scientist would lead the historian of science, seeking the precise antecedents of every step in the progress of knowledge, awarding posthumous credit for whatever can be interpreted as an anticipation of some later truth, or pricking bubble reputations with new documentation.

The professional historian of science resists the charms of this pedantic antiquarianism, and looks for much more subtle influences and trends than can be discovered in a chronicle of inventions and communications. The history of science is to be regarded as a branch of the general history of culture and ideas. The individual scientist is subject to a wide range of philosophical, religious, aesthetic, political, economic and other ideological influences. At the same time, the course of his research is strongly affected by his social position, intellectual resources, emotional needs, and other personal circumstances. The analysis extends to social and psychological factors that are outside the conventional scholarly record.

The biographical slant to the history of science has great attractions, both for its human interest and as an antidote to naive scientism. A masterpiece in this genre – for example, *The Double Helix* – can show the scientist as a real person, the child of his times, fallible, uncertain, living laborious days, falling out with his fellows, vainly seeking glory and all the rest of it. It can illustrate by example all that one might wish to say about the personal dimension of research (§4.2), not only within academic science but also in the wider social context of the R & D system (§5.6).

But this individualist slant often under-emphasizes collective activities and institutional forms. It is only very recently, for example, that academic historians have begun to study learned societies, communication networks, and other social phenomena within the scientific community. The history of technology, with its symbiotic relationship with pure science over the centuries, has been seriously neglected. The historical role of science and technology within society at large has not yet been studied in sufficient detail to confirm or refute some very conjectural interpretations in which class and economic factors are given primacy. The STS theme is by no means satisfactorily established in the history of science, or in parallel disciplines such as economic and social history. Indeed, the struggle to give highbrow metascience a place in academia is closely allied to the movement for similar themes in science education.

Nevertheless, the historical approach to STS studies has immense educational merits. As has already been remarked, the history of science as a social institution, long-lived, yet changing significantly with time, is a major part of the 'message' to be conveyed. And by looking at institutions or concepts or careers or material resources in their more primitive primary forms, one may come to understand them more easily than in sophisticated contemporary manifestations. Thus, for example, the fundamental purpose and social function of a learned society and its scholarly publications can be seen more clearly in the early history of the Royal Society and its *Philosophical Transactions* than it would be in the commercialized, bureaucratic business of a modern scientific society. By its very distance from present-day concerns, the trial of

Galileo illustrates the conflict between critical science and orthodox authority far more clearly than, say, a study of the Oppenheimer or Lysenko affairs.

Romantic distance adds dramatic force. The enlargements of scale, the caricatures of personality, the mythical features that distort our image of historical events often serve to strengthen the lessons they teach. Just as conventional science needs its 'giants' of the past so STS education needs its traditional typologies, venerable institutions, dramatic episodes and heroic figures. What could be a better introduction to the sociology of knowledge, for example, than an account of the famous Oxford meeting of the British Association in 1860, where Darwin's Theory of Evolution was so furiously debated. The conscientious teacher, with the apparatus of exact scholarship at hand, may be well aware that the story being told is not quite so black and white as it may appear; but this is the ineradicable deficiency of all historical teaching, which can never, so to speak, project more than a '1066 and All That' version of what really happened in the past.

But it is easy to underestimate the real difficulty of setting up a satisfactory historical image of the development of scientific ideas. The history of science is much less comprehensible, and much less interesting to the immature student, than it seems to the experienced science teacher. The chronicle of episodes and influences is weighed down with detail. Who now cares which of the twelve possible claimants should be considered the first to discover the Law of the Conservation of Energy? What does it matter that Leonardo's anatomical drawings were lost, and Mendel's paper was ignored? Conflicts over priorities are 'far off things and battle long ago'. Science students are strongly tinged with the fundamentally anti-historical attitude implicit in successful research – that the follies and errors of the past have at last been transcended, and the memory of them is useless baggage, safely to be jettisoned.

The academic guardians of the history of science do not always help to make their discipline familiar and popular. Their research is often concerned with very subtle issues, from which most scientists are cut off by their lack of general education. It is all very well to interpret the Scientific Revolution of the 17th century as an outcome of the Renaissance and the Reformation – provided that your students have a good idea what these cultural climacterics themselves portended. Most of them will not have the foggiest notion what the pre-Socratics had in it for Socrates, nor the neo-Platonists out of Plato, nor why logic is Aristotelian. The history of science is a lovely academic discipline, but mainly for the *aficionado*.

The skilful teacher with a clear idea of the object of the lesson can avoid these traps and thickets, and make a great deal of progress by the historical approach to STS themes. But the objectives of the STS movement in science education are not automatically to be reached by teaching the history of science to science students. The truth is that science has many different 'histories' – individualist, technocratic, philosophical, sociological, ideological, etc., according to inclination. It is the responsibility of the STS teacher to make such use of these diverse slants on the subject as is appropriate, and not to assume that any one of them is a correct view. The history of science is not, of itself, an 'approach' to STS education: it is a method, or a mode by which very important STS themes can be enlarged or illustrated.

7.6 The philosophical approach

To the academic mind the essence of a subject is its deep theory. Chemistry, we say, is the outward manifestation of the valency bonds between atoms; biology only makes sense when we have deciphered the genetic code. If our subject is to be science itself, then it should be studied in the light of *its* theory, whatever that may be. It would be a betrayal of our commitment to disciplined inquiry if we did not teach about science in that spirit. For many scholars this seems the only legitimate *academic* approach to science studies.

The notion of *metascience* – a theoretical discipline concerned with the question 'What is science?' – is central to the whole STS theme. If such a discipline genuinely exists then it is clearly our duty to communicate its findings and to establish it as the foundation of STS education.

Conventional science education gives a stock answer to this question (§3.3): science arises by applying the scientific method, whose metascientific justification is to be found in *philosophy*. Although working scientists are seldom well-informed about the philosophy of science, and are usually very sceptical about its practical value, they often feel that it would be a good thing if science students were taught a little bit about it, and are willing to include it as a subsidiary topic in many courses of 'valid' science.

This point of view is now well-established in science education. In many universities and polytechnics there are small departments of History and Philosophy of Science, encouraged to pursue their own specialized research interests in return for elementary service courses for the mass of ordinary students of science and technology. In the opinion of orthodox science teachers of a liberal persuasion, such courses tell all that need be said *about* science in a formal academic context.

For supporters of STS education, a philosophical approach to science studies has considerable tactical advantages. It seems much easier to exploit this conventional opening towards a broader curriculum than to break out of 'valid' science in other directions. From an intellectual point of view, also, it seems entirely appropriate. Science is quite clearly an organized body of knowledge, publicly accessible in its archival literature and pre-eminently rational and objective (§4.5). These qualities are open to logical analysis and criticism according to the best traditions of academic philosophy. It is a commonplace, moreover, that many of the questions to which there are now scientific answers were once the concern of philosophers, and that many modern scientific theories raise subtle philosophical issues which are still unresolved. There is such continuity across the centuries and across disciplinary boundaries that the history of science itself can be written as if it were no more than a child of philosophy grown larger and lustier than its embarrassed parent. An approach to metascience through philosophy can draw upon an ancient academic discipline, for established knowledge, for firm foundations of principle, for research programmes, for theoretical insights, for appropriate teaching material and for experienced teachers.

But how far does teaching the philosophy of science contribute towards the goals of STS education? It is quite essential that science students should have a clear idea of what distinguishes scientific knowledge from other claimants to belief. The epistemological challenge – to what extent, and why, should science be believed – is fundamental. One of the major objectives of the STS movement is to escape from a supposed choice between a scientism that insists that *all* science, and *only* science, is perfectly O.K. and an antiscientism that makes all scientific knowledge entirely questionable or reprehensible. The naivety of this dilemma can only be demonstrated by a study of the intellectual characteristics of scientific knowledge, of the criteria for scientific validity, of the status of various branches of science, of the roles of observation and experiment, conjecture and theory, argument and proof. Some of these points can be illustrated very vividly by examples drawn from a particular branch of science as it is being taught, but it is part of the message of interdisciplinarity in research that they should be shown to be quite general, both for basic knowledge and for its applications.

Unfortunately, the conventional curriculum of the philosophy of science does not necessarily give very adequate instruction on such issues. Although it has broken down its positivist, inductionist fences, to move into much more open territory, it still tends to cluster very heavily round a certain number of traditional problems which have little relevance to research in action. Much more attention is given, for example, to the status of hypothetical entities such as the mathematical constructs of theoretical physics, than to the methodological difficulties of less exact sciences such as psychology. The analysis is often very narrowly conceived within a limited axiomatic framework where logically idealized proofs can be made much more convincing than in more loosely structured theoretical systems. The philosophy of science easily tends towards a more sophisticated modification of naive scientism where physics ('natural philosophy') still rules supreme.

Indeed, for an STS teacher who has become familiar with other aspects of the model of science presented in previous chapters there can be no going back to a purely philosophical characterization of science. The metascientific foundations of this model are as much psychological and sociological as philosophical in the accepted sense. Any approach to STS education through the philosophy of science needs to be broadened very quickly, and driven forcefully across some well-defended frontiers into these other disciplines, if it is not to get bogged down in abstractions and argumentation. Nor does academic philosophy have much to say about the more technological aspects of R & D, until one reaches the practical interface with society where issues of moral responsibility become salient.

STS education must be built upon a theory of the nature of science. There is much to be said for concentrating initially on a point near the traditional centre of science studies – the philosophical analysis of well-established scientific theories – before trying to broaden the discussion towards more social aspects. Unfortunately, the philosophy of science is quite a difficult subject, which is not easily adapted to the overall needs of education. A very small proportion of science students find great satisfaction in its rigorous, logical, sceptical approach to fundamental questions of existence and reality, certainty and doubt. But the great majority are not very reflective about such matters, and regard the refined pleasures of philosophical debate as so much sterile logic chopping, divorced from serious practicalities. If such an approach to STS education is to be attempted, these issues must be presented in simple, direct language, without too much respect for scholarly niceties, and shown to be relevant to the solution of real problems in the real world.

7.7 The sociological approach

The current trend in science studies is away from philosophy towards a more sociological approach. Contemporary models of the research process lay stress on institutions and communities, within academic science (§4.4) and in the R & D system as a whole (§5.6). STS themes are as much about society in general as about science or technology. In the opinion of many scholars, sociology has replaced philosophy as the fundamental metascientific discipline.

From all that has been said in previous chapters, this seems a very reasonable point of view. There is little doubt that the traditional philosophy of science has given too much weight to the cognitive and personal dimensions of the subject, and has largely ignored its collective phenomenology and its social consequences. An approach from the general direction of the social sciences leads at once to some of the most important components and connections of the R & D system. From precise investigations of the internal structure of an invisible college through to the most tenuous commentary on the sociology of knowledge, from the moral dilemmas of the scientific expert to decision-making in the R & D bureaucracy, sociological, economic, political, and psychological issues permeate the whole STS theme.

Teaching about modern science calls for a large amount of social thinking. At every level it must be concerned with the behaviour of human groups, from research teams to national governments. Some of our thinking on such matters can still be carried on in the traditional language of human affairs – 'leadership', 'loyalty', 'teamwork', 'conflict', etc. But it is impossible nowadays to talk seriously about such things without using concepts such as 'norm', 'role', 'ideology' or 'goal displacement' drawn from academic sociology and its cognate disciplines. Nor is it possible to follow a great deal of the more advanced scholarly research in the social studies of science without a thorough formal grounding in such disciplines.

It is natural, therefore, to propose that STS education should start with an elementary introduction to the sociological concepts that are relevant to the research process, and that the R & D system should then be studied primarily from a sociological point of view. If the basic theory of science is essentially sociological, it can only be properly presented in the language and categories of that discipline. If students are to understand the true nature of science, then they must be given access to insights that can only be formulated in this more rigorous language. In the past we have turned to experts in the philosophy of science for authoritative answers to our metascientific questions; now, it seems, we should enrich the education of science students by passing them over to the sociologists of science for yet deeper answers.

The sociological approach to STS education is thoroughly justifiable in abstract principle – but very difficult to follow in practice. A natural bent towards the sociological mode of thought is even rarer amongst would-be scientists and technologists than a taste for philosophy. The vocabulary, the concepts and the research results of the sociology of science seem just as abstracted from reality or distant from ordinary life as those of the philosophers. There is the same irresistible tendency amongst the academics to drag the student into the more recondite current controversies – how much weight should be given to the Mertonian norms; how directly are research programmes governed by external social forces; can an invisible college be defined by reference to co-citation networks – at the expense of an elementary account of the way things are.

Indeed, some of these controversies are so sophisticated, so value-laden, and ultimately so unresolvable, that they can easily carry the naive student off into quite absurd positions, such as that scientific knowledge is all very relative and questionable (relative to what other knowledge, pray – and questionable by whom, saving your presence?) Such positions, like philosophical solipsism and total scepticism, are instructive because they are impregnable to formal argument, but when they are expounded into an intellectual vacuum they can be just as silly and damaging as their positivist antitheses. Instead of combating the doctrinaire scientism of conventional science education, the sociological approach may convert a few students to an equally dogmatic religion in the opposite direction. From an educational point of view, it may not be a bad thing to challenge the student's simple faith in science and personal commitment to a technical career, but not by pushing him or her through a course that seems to teach precisely the opposite of all that has previously been learnt.

This sort of curriculum can also have the effect of alienating the STS movement from the main stream of education in 'valid' science and technology. This may seem an unworthy, purely opportunist objection. But most natural scientists are somewhat sceptical of the intellectual claims of sociology. Many wise, thoughtful and open-minded scholars, teachers, and men and women of action still harbour grave suspicions of the theoretical validity and/or practical utility of the more general themata of this discipline. They would not deny the value of the statistical data, quantitative information, descriptive analyses and interpretative hypotheses that are brought together by research in this field. Nevertheless, they have good reason to insist that these do not yet combine into deep theories of science, or technology, or of society at large. In their experience, it is less damaging to teach a subject empirically and phenomenologically than to impress upon it a theoretical framework that has not been thoroughly tested and proved. They would argue, in the name of academic freedom and pluralism, that it would be wrong to cast STS education into a sociological mould whose paradigms are no more satisfactorily established than those of the naive rationalism and positivism of the older philosophical approach. This danger of the sociological approach should not be treated lightly.

7.8 The problematic approach

The world, alas, is full of problems – energy and population crises, the exhaustion of resources, environmental pollution, famine, pestilence, war and the threat of war. Even if we do not belong to the Doomsday party in the debate on *Limits to Growth*, we must all feel concern about these apparently incurable symptoms of the world *problematique*.

Science and technology are powerful factors in world affairs. Scientists and technologists are bound to be deeply involved in producing some of these problems, and perhaps solving some of them. It is appropriate that the prospect of such involvement should be discussed in the course of their training.

Yet conventional science education (§3.7) completely ignores this aspect of the human condition. The implication is that all such matters can be left to the political and administrative authorities (§6.4), or that one could learn all about them in due course, if the occasion should so demand. This cool disregard for the great issues by which we are surrounded now seems morally indefensible. The STS movement owes much to a deep concern to demonstrate the connection between the growth of scientific knowledge and the growing problems of the world at large.

For many STS teachers, therefore, it seems obvious that the new curriculum should begin with an account of the *problematique*. Since this is seen as the central social issue of our times, it is the natural starting point, within 'society', for a study of the social role of science and technology. Young people may sometimes seem not to give much thought to their own personal careers, but they are not unmoved by news of abominable human conditions in other countries, nor by confident predictions of grave disasters in their own lifetimes. Conscientious teachers respond to this concern, and seek to prepare their pupils to meet or overcome these horrors. This approach to STS education thus has great emotional force behind it.

From a pedagogical standpoint, also, it has great merits. It is the natural antidote to the abstractions and academicism of philosophical and sociological metascience. By choosing one or more of the problematic issues, it is possible to illustrate the social role of science and the way it actually works, both under external influences and in its internal functions. By reference to concrete examples which are of deep concern to any civilized person, such a course can acquire an immediacy and realism that is difficult to achieve in other ways. By opening up direct channels between the research findings of basic sciences such as biology, and poignant problems of humanity such as malnutrition, this approach to STS education has the power to release quite new forces of commitment and social responsibility amongst scientists and technologists – and new factors of directed enquiry and rational action in the social and political sphere.

But it calls for just as much wisdom, discretion, and breadth of vision on the part of the teacher as any other approach to the STS theme. Despite its many virtues in principle and practice, the way to STS education from society and its problems has to be followed cautiously, with a clear eye for some major pitfalls. Attention must be fixed on the primary objective – to teach about science and technology in relation to society – and not simply about science-related social issues.

The problematic theme unfortunately tends to drift away from this objective towards much wider and more distant horizons. Neither the great social issues of today nor the prospective problems of the future are exclusively technological or scientific in origin or in the means by which they might be tackled (§6.2). The Energy Crisis, for example, has commercial, industrial, legal, administrative, fiscal, syndical, national, military, ethical, medical and aesthetic aspects, all of which would need to be taken into account in serious policy-making. What a science student would regard as the significant technical factors, such as the geological location of fuel reserves, the constraints imposed by the Second Law of Thermodynamics, or the prospects for the exploitation of nuclear fusion, are only a fraction of the variables in the equation. An STS course that is deliberately centred on such a theme must, in all conscience, demonstrate the relevance of this multitude of considerations, and hence must rove very widely over the whole political, economic and social scene.

Most supporters of the STS movement favour this trend towards a very broad general education centred on social problems. It is conceived as a means of enlarging and enriching the upbringing of students of science and technology, and sensitizing them to the social dilemmas, political conflicts, human values and moral imperatives of the real world. To grasp the significance of the world *problematique* and to be equipped with the knowledge needed to deal with it is regarded as a liberal education in itself.

But STS education cannot carry such a heavy load of liberal conscientiousness on its own back. In trying to cover every aspect of human affairs, it can give no more than a superficial glance at all sorts of deep and difficult questions (§5.3). It can easily lead to a ludicrous situation in which STS enthusiasts are expatiating amateurishly on matters which are already being taught much more competently under more conventional headings in Departments of Economics, Politics, Geography and History. Under the worst conditions, STS courses may even fall under the influence of some simplistic but all-embracing formula, such as that all the troubles of the world could be solved if only we returned to *laissez-faire* economics, or that they can all be blamed on the wickedness of the bourgeoisie. The good intentions of the STS movement are no excuse for taking up doctrinaire positions that are just as blinkered as the scientism that we are endeavouring to combat (§7.1).

One of the pitfalls of the problematic approach is that it may give a very misleading account of the R & D system. It is easy, for example, to overemphasize the populist, antiscientific opinion that science 'got us into our present mess' because it is out of social control, and is governed only by the irresponsible search for knowledge for its own sake (cf. §3.8). Or there may be an unexpected and almost unconscious return to the technocratic fold, where science is called upon to 'get us out of this mess' as if it were the only rational instrument for the diagnosis and cure of social ills (§3.5).

If the problematic approach leaves the student in either of these grossly oversimplified attitudes towards science, then it will be more damaging than traditional science education which only refers to such matters very obliquely. The whole burden of our argument (§6.2) is that the R & D system is neither an all-powerful instrument by which our every desire can be achieved by pressing the appropriate buttons, nor is it a blind monster blundering around in society like a bull in a china shop. Scientistic and antiscientific opinions are merely the most elementary terms in a much more complicated line of argument where a multitude of factors of freedom and responsibility, institutional independence and social control, foresight and serendipity, calculated profit and wild speculation, individuality and communality must be combined. Discussion of the world *problematique*, in whole or in part, is an excellent starting point for STS education; it must, nevertheless penetrate deeply enough into the R & D system to uncover its structure and functioning in the various dimensions of knowledge, personality, and social interaction.

7.9 What is the best approach?

There are numerous intellectual approaches to science, technology, and society. The seven different ways that have been discussed here do not exhaust the possibilities. Many other approaches can easily be imagined. None of them is necessarily exclusive of the others. In practice, these purely theoretical themes overlap and combine inextricably. The problematic approach, for example, is interdisciplinary by its very nature; and it would take peculiar blindness and ignorance to keep a touch of history out of an STS curriculum.

And for every aspect of the subject, a variety of pedagogic techniques are possible. Some would teach didactically, in formal lessons or lectures. Others adopt the Socratic method, continually drawing out student thought and opinion by careful questioning. Theoretical principles can be introduced from the start, and then illustrated by examples, or can be uncovered along the way through studies of particular cases. Each subject can be organized around a series of project topics, which students can be made to investigate in small groups. There are obvious occasions for reference to books, journals, newspapers or less public sources of information. Students may be encouraged to undertake original work on their own account, from essays on historical topics drawn mainly from standard works of reference to experimental research in support of environmental conservation or industrial safety.

Each intellectual approach, each pedagogic technique, each style of lesson, lecture, seminar, case study, essay topic or research project has its characteristic merits and demerits. It is not the place here to suggest to any teacher how to do his or her job. Academic freedom puts the responsibility clearly on each teacher or group of teachers to construct or choose a curriculum and to set about teaching it in the manner most appropriate to the situation.

But that responsibility cannot be exercised without reference to experience. As I have already remarked (§7.1), the most agreeable characteristic of STS education is that it is not sharply constrained by intellectual necessities or pedagogic traditions. There is no need, as in teaching a foreign language, to make the students learn a large amount of finicky detail. The subject is not, like mathematics or physics, a hierarchy of structured concepts to be climbed in proper order. There are no disciplinary frontiers, as between political science and sociology, to be respected or disputed over. The STS theme has both descriptive and analytical aspects, it is open ended, it can arouse interest and feeling, it can exercise hard thinking and thoughtful action. Almost any way one starts into it, there are doors and windows into whole palaces of new understanding.

This absence of formal constraints and academic traditions has proved altogether too seductive. Enthusiastic teachers have plunged into new courses, with unfamiliar subject matter to be taught by untried methods, apparently heedless of the difficulties they were likely to encounter. Some of these courses have proved very successful; others have been almost complete failures. What have we learnt from this experience? Is there an ideal approach to STS education?

Nobody has yet undertaken the very considerable task of collecting and collating the empirical evidence by which such a question could be answered. Even if the experience of the STS movement in the past ten years could be codified, there could be no consensus about how it should be assessed. Some would give weight to one aspect of the subject, some to another. As has been emphasized in earlier chapters, there is still much confusion and little agreement about the objectives of this educational innovation, let alone its methodology.

In a pluralistic society, where education is entrusted to a large number of more or less independent institutions, this is not an unhealthy situation. The way to curb totally idiosyncratic anarchism or parochial ignorance and incompetence is not to attempt to define and impose a single curriculum, but to encourage discussion, debate, exchange of experience, openness to external inspection and validation, collective initiatives, etc., throughout the STS movement. Many such channels of communication and mutual criticism already exist, although there is notable weakness at the centre, where, for example, there is no organized forum, no meeting place for the invisible college of STS education at all levels (see chapter 9).

What experience has taught us, however, is that there is no *ideal* approach to the STS theme. Each of the approaches discussed in this chapter has its own convincing rationale, its own pedagogic advantages, its own characteristic range of validity – and its own deep pitfalls. Most of these pitfalls are encountered blindly, in an excess of zeal. Having adopted a particular approach, the teacher tends to focus narrowly upon the corresponding aspect of the subject, distorting the image of science presented to students or losing their attention in a forest of academic detail. By its very nature, the STS theme is eclectic and ecumenical; it can never be conveyed adequately along a single dimension, whether of rationality, personality, social interaction, economic necessity or political power. It is as much contemporary as historical, as much psychological as sociological, as much practical as theoretical, as much individualistic as ideological, as much specialized as generalizable. The teacher who fails to realize that it is a many-sided subject cannot deal fairly with his or her pupils.

There is no single best approach. There is not even an optimum recipe for combining the various aspects of the STS theme – mix six ounces of History with three tablespoons of undiluted Philosophy and a pinch of Sociology, season with Relevant Problems and bake for three periods a week in an Interdisciplinary oven at a moderate Ideological temperature! Teachers must make their own lists of ingredients, and learn to combine and cook them to suit the tastes and nutritional needs of those to whom the dish is to be served.

And that is where we must now return – to the diverse educational needs and abilities of a very wide range of young people, at various ages, in various stages of education, making their way towards various careers and callings. How should we be thinking about what they should be taught about science in its social context?

The major constraint on STS education is that it cannot be imposed as a formal discipline, as if it were an essential foundation for a learned profession or practical craft. It cannot be justified by strict vocational necessity, like the study of anatomy for medical students or of mathematics for engineers. It takes its place in the school or college curriculum as a minor topic, a voluntary option, an item of General Studies. It is usually given little weight in entrance examinations, and may count very little towards degree results. It cannot compete directly with the conventional scientific disciplines for teaching time – and hence for academic resources and respectability – in the traditional, unquestioned terms of intellectual 'validity'.

In the long run, STS education can only justify itself by its success in holding the attention and interest of students, and making them a little wiser and a little better in recognisable ways. In an ideal world, the contents and style of teaching would be tailored to the personal needs of every single pupil – as it might be in an educational system modelled on Oxbridge tutorials. In practice, it calls for serious thought about the characteristic capabilities, tastes, and educational goals of the various groups of students in schools and colleges where STS education is now being undertaken.

8

Enlarging science education

8.1 Matching diverse needs

Conventional science teaching is built around the platonic ideal of a *total curriculum*. This is simply all that is known, or all that one might think worth knowing, about the subject in question – chemistry, botany, invertebrate palaeontology, or whatever it is. Within this curriculum, each topic, concept or technique can supposedly be given a characteristic meter reading along the scale from elementary to advanced. When account has been taken of prerequisite knowledge of other topics (§2.1), such as the calculus that will be needed for classical mechanics, or the organic chemistry required for molecular biology, the possible courses of study leading to the research frontier practically define themselves.

As we saw in chapter 1, the system of science education is dominated by this image of its fundamental purpose. The actual contents of most science courses is largely determined by the needs of the relatively small proportion of students who are hoping to proceed to the next level up the pyramid of valid knowledge. Although this means that the precise educational needs of the majority of students are not taken directly into account, it is hard to argue radically against the principle that it is better for them to learn what can be 'validly' taught at this level than to set out towards much less definite goals. Educational development in the sciences and technologies has traditionally been much more concerned with pedagogic technique and styles of exposition than with the topics to be taught and the order in which they are presented.

A fundamental difficulty with STS education is that the whole notion of a total curriculum is inappropriate. As must have been clear from chapters 4–6, science spreads without frontiers into technology, which is connected with every aspect of cultural, intellectual and practical life. The deep theory of the research process is not so highly structured or soundly established to define a necessary central core around which all the rest can be arranged. Although a basic model of the R & D system (chapter 5) should be regarded as fundamental to science studies in general, there is no metascientific theory that is so well established, and precisely analysed in its internal relations, that it should be made the final goal or explicit theme of STS courses. As we saw in chapter 7, the social role of science can be approached from many different directions, at various levels of sophistication, without significant constraints of prerequisite knowledge. It is perfectly feasible, for example, to enter into a philosophical discussion of scientific method (§7.6) without much formal knowledge of philosophy, or to teach at least the internal history of a branch of science (§7.5) with little reference to general history.

School and college courses in science, technology and society are not, therefore, strongly determined by familiar academic disciplinary considerations. As must continually be emphasized, they must be matched, rather, to student needs and interests. At every level, in every educational institution, there is a tug of love between the relatively unacademic prospects of the majority of the student group, for whom this is likely to be the terminal contact with formal education (chapter 1), and the more intellectualized attitudes of the ablest students who are climbing to yet another step up the academic ladder. In conventional science education, this tension is almost always resolved in favour of the academic élite (§2.2); but in STS education, students are seldom being taught specifically for entry to a more advanced level in the same subject, and have no reason to pretend that they are working their way up some pyramid of valid scholarship to an apex of learning where all will be revealed. In this field of education, the real needs of the majority of students in each course must be sincerely respected.

What we must do, therefore, is to consider the specific educational objectives of the major groups within science education, and try to show how these needs might best be met by STS courses. It is necessary to relate the contents and approach of each such course to the likely futures of the students concerned (chapter 1). Is it simply a contribution to their general competence as citizens of a civilization where science and technology raise fundamental problems? Is it a process of liberal intellectual enlightenment, providing them with an enlarged framework of social, ethical and philosophical ideas? Or is it intended to clarify and rehearse some of the dilemmas that they may have to face in the course of specialist careers in science and technology? All three objectives may be thought appropriate, although with very different weights in different cases.

The real difficulty is not to think up all sorts of questions that might be asked about STS curricula in general: it is to think through the answers to such questions in relation to particular courses. At this stage in this book, it would seem almost inevitable that we should begin to apply our abstract arguments and theoretical principles to the actual courses being taught in various schools and colleges in the UK and other countries. Surely we should now be turning from the heady idealizations of educational theory to the sober realities of educational practice.

Unfortunately, it is just here, where I might have been expected at last to get down to the brass tacks of real examples of satisfactory STS curricula, that I feel I must disappoint the reader. It is certainly not because I believe that such curricula do not exist, for I am acquainted with a number of them in a superficial way. I have also come across a number of other STS courses which seem gravely defective in ways that could be cured with a little elementary thought along the lines of this book. But the total number and diversity of all such courses is already so large that it would be grossly unjust to make such evaluatory choices, and to distribute kicks and ha'ppence within a band of sincere and devoted teachers, without the much more detailed information that might give weight to such opinions. Such a research project would have to be so serious, so expensive, and so lengthy that it would delay the completion of this present book far beyond any useful date of publication. It is precisely from such motives of scholarly scrupulosity that I have decided not to name specific courses in what follows, even though the general line of argument is based on some empirical observations of actual practice.

The various types of STS course that will now be considered are no more than broad schematic categories. Any simple classification of the many branches of science education is bound to be inaccurate in detail. Although school and college students can easily be grouped by age, by the formal qualifications they are seeking, and by crude vocational intentions, there are more subtle subdivisions that may be just as significant in the design of an STS course. It is difficult to make full allowance for the different approaches that might be thought best for students in different scientific disciplines or preparing for different technological professions. Institutional conditions may be equally significant. Teaching science in the Sixth Form of a single sex private school to middle class children confidently bound for managerial or professional careers is not at all the same as it is in a mixed-up comprehensive school in the heart of a big city, where the background experience of the pupils is much more cramped and disheartening. All these very important distinctions, which must surely be recognized by the teachers involved, will have to be ignored in this oversimplified analysis.

8.2 General science for the general public

Since STS studies are not yet supporting an academic apex reaching for the heavens of research, we can start rationally from the broad base of general education. The school curriculum up to 16+, when most children leave formal education and seek more gainful employment, contains quite a lot of science. In the later stages this is directed towards the certificate examinations of CSE and GCE O level; but for the majority of pupils it is not being learnt as a prerequisite for a technical job or for entry into a more advanced course of study. The prime purpose of science education in the pre-certificate years can only be to prepare people for ordinary life in our scientifically oriented, technologically advanced culture.

Until quite recently, the role of science as a major component of general education was interpreted in a very stilted and traditional manner. The conventional disciplines – Mathematics, Physics, Chemistry and Biology – were taught separately along rigid lines leading up to the certificate examinations and beyond. This tradition still persists in many secondary schools, although the intellectual objectives, pedagogic techniques and detailed syllabuses of school science courses have been revised and reformulated from time to time – often to meet new demands from academia.

At the same time, as the academic and non-academic streams of secondary education have been merged in the comprehensive school, science teaching in the lower and middle school has been set free from the absolute tyranny of highly 'valid', scholastically pure and rigorously prescribed examination syllabuses. Schools and teachers now have considerable liberty in working out new curricula and teaching by more imaginative methods towards much less academic goals than in the past. In fact, science education for the general public is going through a period of rapid change, where it sometimes seems that it is permissible to introduce almost any new material or any new method of teaching that teachers can think of and their pupils can master or tolerate.

For the supporters of STS education, the most significant feature of this transformation is the breakdown of formal barriers between the traditional science subjects. In most comprehensive schools, these subjects are no longer taught separately until a year or two before the certificate examinations. There are even moves to extend these integrated science courses right through the middle school, up to appropriate examinations for CSE and O level. In other words, the interdisciplinary approach to STS education lies wide open.

The interdisciplinary approach (§7.4) stresses the practical relevance of scientific knowledge to the technical problems of society. To make the most of the opportunity, it needs to be accompanied by an up-to-date attitude towards the more conventional science subjects, demonstrating their intimate connection with advanced technology. This attitude is favoured by modern-minded teachers and in society at large, so that most of the new integrated science curricula tend strongly in that direction.

Is this approach, stressing the practical role of science in society, a satisfactory introduction to STS themes in the middle school? Some enthusiasts think not. They argue for a completely new science curriculum, in which philosophical (§7.6), sociological (§7.7) and problematic (§7.8) topics should replace a significant proportion of the subject matter of 'valid' science and its technological applications.

This argument has considerable force. The middle school is the terminal contact of most people with science education. Surely they should be armed against the dangers and follies of our modern age by being taught something about the scientific method, about the limited range and uncertain validity of scientific expertise (§6.3), about the deceits of commercial and ideological distortion of research, about the risks and benefits of technological innovation (§5.3), and so on. The present book is based upon a firm belief that such matters should be better understood and more widely discussed than ever before, at all levels of political and social influence.

But can this sort of thing be *taught* at the very grass roots of schooling, just as the green tips of the new blades are peeping through the ground? As we saw in chapter 7, philosophical and sociological metascience is very intellectually demanding, especially for young people with a very restricted experience of science or of the world. Even at college level, it only appeals to a minority of science students; can it ever be made attractive, or even comprehensible, to the vast majority of school children at 15 or 16+? For adults, the problematic approach appears more intelligible and interesting, but it presupposes a degree of social concern and 'political' maturity that is just not to be found in most early teenagers. The main virtue of this approach to STS themes is that they can attach themselves to such concerns and provide an ideal vehicle for their further expression; the notion of reorienting science education in this direction is likely to prove premature for the average young person at this stage in his or her intellectual and emotional development.

Nevertheless, much can be achieved as science teaching moves towards interdisciplinary relevance at this formative stage. There are opportunities to depict scientific knowledge in its social context, and to show that science is neither a collection of useless facts and mysterious theories, nor yet a box of tricks for getting things done, regardless of the consequences. Ways can be opened to other school subjects, stressing the links of science with all aspects of our culture. Science teaching needs to be connected with the humanities, for example, by reference to the historical importance of scientific and technological revolutions, and with environmental subjects, such as geography, which already contain many scientific components and which deal directly with some of the major problematic issues of our times, such as the exhaustion of mineral resources and the expansion of populations.

The current ferment of innovation in basic science teaching is anarchical and incoherent. Beyond the call for social relevance and transdisciplinary integration, it lacks a detailed programme. In some ways, the most exciting challenge to the STS movement for the next few years may be to catalyse this process of transformation, and provide it with a more definite rationale.

8.3 General Studies for the science specialist

In the Sixth Form, the interdisciplinary permissiveness of earlier years gives way to the 'valid' science subjects of the GCE A level. Those pupils who have survived through the school-leaving age are mostly heading towards more skilled employment, in technical work and apprenticeships (§1.4) or more remote professions beyond higher education. In the English educational tradition, they must now become much more selective in their studies, seeking good grades in two or three (occasionally four) subjects. Although, in fact, most of them have no precise vocation clearly in mind, they tend to choose subjects within a narrow range – physics, chemistry and mathematics, say, or chemistry, biology and home economics. Not all those who study A-level science subjects already regard themselves as 'scientists': generally speaking, however, we are now dealing with young people who are likely to become closely involved with science and technology, in one way or another, as researchers, technicians, engineers, doctors, medical auxiliaries, teachers, or industrial managers, for the rest of their lives. For these, at least, there can be no doubt of the importance of STS education. But how is this to be brought into the Sixth Form curriculum?

It is easy to insist that their A-level courses in the various branches of 'valid' science should be made as relevant as possible. As we saw in the previous section, there are many little ways in which this can be done without seriously interfering with the progress of the student up the conceptual hierarchy of the subject. It is only stale tradition, for example, that illustrates a new scientific concept by reference to an antiquated experimental set-up rather than to a contemporary technological application – Boyle's Law, say, by reference to a bubble of gas trapped over mercury in a glass tube, rather than to deep sea diving and weather forecasting.

But the ideal of teaching integrated, transdisciplinary science is Utopian – not so much for any logical reasons about scientific knowledge itself, but because of the limitations of teachers and the inertia of the examination system for entry into higher education. And the approach to STS education through interdisciplinarity (§7.4) and relevance (§7.2) really calls for much more time and effort than can reasonably be spared for it in a course that is primarily devoted to a 'valid' science subject. The tradition of high academic standards in A-level science is very strong, and in most respects thoroughly justified. If, at the age of 17 or 18, you claim that you are studying chemistry, or botany, then you ought to be doing it properly. There is an immense amount to learn, and an immense amount to understand. The standard of achievement required to get a high grade in the examination is not arbitrary: it represents just what can be done by able students who have worked hard under the instruction of competent teachers. So although we might make a strong case for the inclusion of some socially relevant topic, this will have to compete for time and attention in the syllabus with many other topics that can easily be justified for their scientific validity, for their intellectual subtlety, and for other traditional virtues (§2.6).

This is a highly significant point for almost all that follows in this chapter. The plea for STS education cannot be satisfied simply by making the conventional topics of science education somewhat more relevant to technological and social issues. This is, in itself, a highly desirable trend which would not be opposed by many orthodox science teachers. But it must not, it cannot, be pressed so far as to compromise seriously the scholarly standards of valid science. STS education is bound to be defeated in a direct confrontation with the whole educational apparatus of conventional science and technology; if it is to establish itself it must present itself in a complementary role, within or parallel to that system.

There is thus little hope of achieving the objectives of the STS movement simply by pressing for a thorough reform in the teaching of ordinary science at A level and beyond. The real need is for distinct courses of study specifically directed to STS themes. What approach should be adopted in such a course for Sixth Form 'scientists'?

The case for a vocational approach, as discussed in §7.3, is not convincing. It is true that these students may be more definitely committed to scientific careers than at earlier stages in their school lives. A number of quite specifically vocational subjects may be taken up at this stage. But most of these are taught as foundations for much more professional training later – and it is there that the vocational aspects of STS education should be emphasized. Thus, the place to study the ethical dilemmas and social role of medicine and its cognate professions is in the medical school itself, where the student is much closer to the realities of the job. For those technical workers whose formal education is likely in any case to terminate at 18+, a vocational approach would be much too restricted in scope. Suppose, for example, one were designing an STS course for an apprentice electrician on an ONC course at a Technical College: how would that be different from what would be suitable for an engineering draughtsman, a prospective aircraft pilot, or even an insurance salesman or supermarket manager? In other words, the prime goal of STS studies at this level is to contribute to general education, and to individual intellectual and moral development, and hence to general social competence and civic virtue.

As we shall see (§8.8), there is a trend within the STS movement to press for Science Studies as a distinct academic discipline in its own right. Those who favour this trend anticipate the setting up of an A-level subject syllabus, which they would hope to make comparable in scholarly standard and competitive validity with Physics, or French, or Engineering Technology. They would be bound to argue for a substantial component of metascientific theory – or at least for some of the basic elements of the history, philosophy and sociology of science.

From what has been said above, in the previous section and in §7.6, this proposal is well meant in principle, but misguided in practice. Very few school children are seriously interested in the history or philosophy of science. Very few have enough knowledge of science to appreciate the real point of such studies. Very few teachers know enough about it to teach it properly. And in any case, the subject itself is in such disarray theoretically that a formal course at this level is likely to be either misleadingly doctrinaire or confusedly superficial. Do we want a soft option that would tempt many students away from broadening their scientific education with at least another conventional science subject – or a tough little specialty that is chosen by only a very few eccentric students pushing towards somewhat idiosyncratic academic careers?

The proper place for STS education in the Sixth Form curriculum is under the heading 'General Studies'. This is often an impromptu, unscripted activity, whose contents and purpose vary confusingly from school to school. The whole question of the continuing existence and status of such studies is uncertain. There are arguments about whether or not this part of the curriculum should be formally examined, and if so what notice should be taken of the results. These issues, in turn, depend upon larger decisions about the future structure of the 16+ and 18+ certificate examinations, over which there seems to be endless chaotic controversy. But as things stand at present, this is where a good STS course can find a place in secondary education, and really show its value.

The opening is certainly there; but the task of devising an appropriate course will tax all the educational insights of the STS movement. On the one hand, we have a cohort of young people, serious enough in their intellectual competence and aspirations, poised to enter into more adult concerns, but with very little knowledge of science itself, nor of the multi-facetted society in which it is embedded. On the other hand, we have a complex of facts, themes and opinions about the most sophisticated cultural institution of our times, with many dimensions of action and being. How are these to be brought together into a significant educational activity?

None of the approaches discussed in chapter 7 can be entirely recommended. In fact, at this introductory stage, the notion of a distinct 'approach' to STS topics may be quite mistaken. It smells a little of the notion of distinct 'disciplines', which we have been trying to drive out of science teaching itself at the most elementary levels. The real need is for an integrated course *about* science in *all* its various aspects – historical, philosophical, cultural, sociological, technological, political, economic, problematic, commercial, beneficial, dangerous, equivocal, ethical, etc. – without specific reference to the academic disciplines by which each particular aspect might be more closely analysed. The focus must be on the actual human institution under observation, not on the various media through which it is being observed.

To put such a course over to students at this age, it must not smell of the candle. There must not be self-conscious reference to learned works, name-dropping of prominent scholars, long-winded technical terminology and the other minutiae of serious scholarship. Nor need there be a great deal of advanced science. Sixth Form scientists include biologists who know only very elementary physics, and physics and chemistry students who have learnt very little biology: the STS course must include both groups, mirroring the unity of science itself.

Indeed one of the major attractions of such a course is that it can easily be made accessible to all Sixth Form pupils, including those specializing in the humanities. Not only is it highly desirable to keep the 'non-scientists' in touch with science: they themselves bring to STS education an essential component of humanistic, historical, socially conscious culture that is often missing amongst science students.

Unfortunately this same humanistic component is not widely cultivated by science teachers. For this reason, the development of STS courses for Sixth Form General Studies cannot be left entirely to the conventional science educationalists, very few of whom have the breadth of knowledge about STS topics to see the subject as a whole and hence to teach about it in an integrated and balanced way. Merely to construct a schedule of suitable topics, and to indicate how they should be taught, calls for a combination of teaching experience, detailed knowledge of various aspects of science studies, and a synoptic vision that is rarely to be found in any single person or small group. The collective labour of developing and testing curricula, constructing examination syllabuses, writing textbooks, and preparing auxiliary material such as student and teacher guides is an obvious priority both for the STS movement and for the whole school science profession in the years to come.

8.4 Enriching the Honours degree

The stairway leads up the academic pyramid, from A levels to the specialist Honours degree. There are several different groups of students in higher education for whom STS education is particularly important. Of these the undergraduate student reading for an Honours degree in a scientific or technological subject is typical.

At first sight, one might be expected to distinguish between the basic sciences, such as chemistry, and technological professions such as medicine or engineering. But for the first year or so at college, all these subjects are taught in much the same manner, with the emphasis mainly on organized, theoretically valid, scientific principles (§1.3). There are big differences, of course, in the details of lectures, classes, laboratory work and practical projects, but this variance can be just as significant between different educational institutions as between 'science' and 'technology'. And as we have already noted, students with basic science degrees often move into technological R & D or industrial management, where their careers may be practically indistinguishable from those with formal engineering qualifications.

The transition from school to college is a great step in any person's life. And yet there is considerable continuity from Sixth Form science into the undergraduate science curriculum. It is quite usual for first year college students to take just the same subjects as at A level – for example, physics, chemistry and mathematics – thus postponing decisive specialization until the second, or even third year of the degree course. There is also the same stress – only more so – on soundly based, rigorous, well-assimilated knowledge and on skilful practical competence.

A degree course in a 'valid' science or advanced technology is intellectually taxing and competitive. There is not much time for cultural digressions and frills. For this reason, the place of STS education in the undergraduate curriculum cannot be vastly different from what it ought to be in the Sixth Form at school. That is to say, it cannot be adequately cultivated by just pushing the teaching of the conventional science topics towards greater relevance, and should claim at least a few per cent of the timetable as a distinct component of general education.

In some quarters, the very idea of enriching a special Honours degree with non-specialist topics is still regarded as heretical. But there has always been a feeling amongst the more enlightened science dons that their students could benefit from contact with a little real culture, and arrangements would often be made for a few extra lectures on uplifting topics, such as anthropology, or Renaissance painting, or the civilization of China. Such initiatives are often very successful as charming interludes. But they fail in their main purpose because they make no contact with the science course itself. This is conventionally aloof from all societal issues: the cultural material chosen to enrich often turns out to be relevant neither to science nor to society, and hence makes no closer connection with the student's educational core than a book he might pick up in a library or a conversation with a student from a different faculty in the Union Bar.

Those science dons who early recognized the need for some actual teaching *about* science turned instinctively to what seemed to be the appropriate academic experts – that is to say, the historians and philosophers of science. In a number of universities, optional courses in 'HPS' are available as subsidiary topics in science degrees. These courses are well established institutionally, but are not usually attended by a large proportion of students of science, engineering or medicine, even at the most elementary level. For a small minority of science students they may prove an important intellectual influence, but they do not significantly enrich the education of the mass of undergraduates. HPS has proved something of a disappointment as an intellectual leaven in science education.

The reasons for this have already been discussed above (§§7.6, 8.3). The metascientific approaches to STS education have only limited appeal, even for students reading for Honours degrees. Scientists and technologists should certainly be made aware of the historical and philosophical attributes of the R & D system, but in its primary academic forms this sort of learning can be very dry, very abstract, and apparently unrelated to scientific and social realities.

What was really missing from the conventional HPS course was any serious reference to the social and technological role of science. This deficiency is not completely remedied by the introduction of sociological topics (§7.7), such as the internal sociology of the scientific community, or grander themes from the sociology of knowledge. It is quite clear now that there must be a much more problematic approach (§7.8), both to arouse the concerns of the student and to illustrate the social context into which science has to fit.

Whether as a revised version of the 'History and Philosophy of Science', or as an entirely new product under a name such as 'Social Studies in Science', a unified STS course that is not closely tied to any particular academic discipline is needed, at undergraduate level, for a wide range of students of science and technology. Like its more elementary counterpart for Sixth Form General Studies, the main purpose of this course should be to present a coherent image of science in all its major aspects – as a technique for acquiring reliable knowledge, as a social institution, as a cultural force, and as an instrument for rational action.

This is not the place to set out a draft curriculum for such a course. Ideally, it would be built around the model of the R & D system described in earlier chapters – but well covered up, not showing the skeleton through the flesh. In style it should move gently from the anecdotal informality of the Sixth Form course towards a position where the points at issue in more advanced approaches could begin to be appreciated. But one would still need to be very careful not to force the scholarly pace, nor to push the student too far into baffling mazes of unresolved academic controversy of little social relevance.

The material for a variety of courses along these lines is now available in abundance. There are a number of books about science and society, about the history of science, and about science policy that are well within the grasp of the average student. They could also read some of the classical papers in the history of science, and should be encouraged to dip into the works of such seminal writers as Thomas Kuhn, Robert Merton, Michael Polanyi, and Karl Popper. More specific educational materials are also available, such as case studies, annotated bibliographies, and curricular guides.

Nevertheless, much of this material is of doubtful quality. The whole field still lacks clarity and coherence, and too high a proportion of what is confidently propounded is really of uncertain validity. That is to say, there is a great deal of hard work still to be done. But the people with the expert knowledge and teaching experience to do this properly are available within the STS community. There is still a big gap between the interests and concerns of the average science student and the research aspirations of the average philosopher, historian or sociologist of science; but this is just the sort of gap that college teachers are trained to bridge. The gulf is not nearly so wide as it is at school level, where STS themes must be translated into much simpler language, and transformed into terms that make sense to the average school child.

8.5 Technology in a social context

The value of a societal component in the training of practising technologists is widely conceded. Industry wants engineers that are not quite unprepared for the commercial pressures under which they will have to work. Political and social critics are unhappy about the ethical priorities of medical practitioners. Nurses, dentists, oil prospecting geophysicists, chemical plant operators, pharmacists, architects, agricultural advisers and a host of other scientific experts are expected nowadays to have social sensibilities as well as technical skills.

It is natural, therefore, to suggest that STS studies should have a strong vocational emphasis for students in all branches of technology as they come near to the end of their professional training. Courses of this kind are now prescribed as part of the formal qualifications for many engineers and are also to be found in many medical curricula. Indeed, they are seriously opposed only in rear-guard actions by the traditional opponents of all theoretical education for technical work, who argue for 'learning by experience' in the societal aspects of professional life just as their predecessors did for the more practical aspects of medicine or engineering.

To be quite fair, opposition to such courses is not entirely reactionary obscurantism. There is genuine doubt whether the arts of industrial management, or the moral dilemmas of medical practice, can be dealt with academically in the same spirit as the design of a dam or the diagnosis of appendicitis. It is not easy for an outsider to decide whether the objection is to the unrealistic pretensions of some of the social sciences, or to the possibility of exposing some of the pretences by which the élite within a closed profession maintain their privileges and authority.

In any case, as we saw in §7.3, a strictly vocational approach to STS studies is quite severely limited in its scope. It can indeed become simply an instrument for reinforcing professional solidarity. Or it can degenerate into a marginal item in the curriculum, giving pedestrian instruction on practical points that could, indeed, be learnt on the job. In the opinion of many STS supporters such courses are scarcely to be counted as contributions to STS education at all.

This opinion is often based upon an ideological prejudice. A vocational course for a student entering an existing profession cannot, after all, be expected to have a very radical or revolutionary flavour. But there is a deeper issue underneath this criticism.

What such courses have tended to lack is any firm basis of knowledge about society and/or science. Students who come upon them in the final stages of a conventional education in 'valid' science and technology have nothing to build on but the implicit technocratic ideology that they have been taking in, unconsciously, for all those previous years. Even the most enlightened teacher of such a course would find it difficult to overcome the political naivety and social immaturity of students whose general education has been so neglected that they simply cannot understand what political, social, economic and moral controversies are really about.

However strong the vocational *justification* for STS studies in the education of advanced technologists, this does not mean that only a vocational *approach* should be used. As was argued above (§§8.3, 8.4), the fundamental tone of all STS education at school, and in the earlier college years, should be transdisciplinary and eclectic, to give a realistic representation of science and technology in all their intellectual and social aspects. Such general courses must be regarded as essential prerequisites for any vocationally oriented course in the final stages of professional training. It is only when students have built up some sort of picture of the R & D system, and of the social context in which it operates, that they can make sense of their own more specialized social role.

Given such preparation, students can undoubtedly benefit greatly from deliberate teaching about the societal and personal aspects of their careers. It is questionable, however, how far this should be very formal and 'disciplined' – whether, for example, it should take the form of instruction on some standard topic in economics, or moral philosophy, or management studies, given by an expert in such matters. It would seem much more appropriate to relate the course to the actual circumstances of professional practice, drawing attention once more to the various aspects and dimensions in which these are to be viewed, and working outwards to more abstract and general concepts. In fact, this is the sort of educational activity in which all members of the department could take part. It is about *their* technology, and they can no more shift responsibility for teaching about its social aspects on to the shoulders of external experts than they can for teaching about its scientific aspects. A modest understanding of the economics of industrial innovation or of the history of mass production techniques should be as much part of the repertoire of a professor of mechanical engineering as theoretical fracture mechanics or the laws of thermodynamics.

Such a course, therefore, cannot be closely confined within a model curriculum. It should be created within a department or faculty, from the interests and accomplishments of its own staff in relation to the prospective careers of their students. Naturally, it should try to be reliable and sound according to the scholarly standards of the social sciences, but not at the expense of immediacy and elementary social insight. For the fundamental purpose of such a course is that it should make explicit, in a more organized form, the type of practical wisdom about people and affairs that has always flowed informally from master to pupil during the transfer of technical skills.

8.6 Preparation for the profession of research

Scientific and technological research can itself be regarded as a profession, for which there should be deliberate preparation in the form of STS courses. As we saw in chapters 5 and 6, the whole trend of the scientific way of life has been towards a much more explicit institutional structure in which bureaucratic procedures have replaced the 'hidden hand' of the invisible college. Even in traditional academic research, there are norms of behaviour, social conventions and ethical dilemmas that may be as unexpected and puzzling as in the practice of architecture or surgery. Surely the aspiring research worker also should be prepared for the psychological and social aspects of his or her vocation?

This issue is dodged in the conventional system of science education because there is no definite point of entry into a research career. On the one hand, a science student about to graduate with a good Honours degree can practically count on entering a full-time postgraduate course for an M.Sc. or Ph.D, or on becoming apprenticed to research in an industrial or government laboratory – and hence is already on the threshold of this profession. On the other hand, even the award of a Ph.D. is not to be relied upon as a sufficient qualification for permanent employment in academic or industrial research. There is often a long period during which the young research scientist is actively involved in the work, and participating fully in its social aspects, before this vocational role is permanently assured. During this period, the norms and conventions of the scientific community are learnt by bitter experience, or by anecdote, or by informal advice from more senior colleagues.

It would be a great mistake to insist that this transfer of research skills and traditions from generation to generation should be made much more formal – for example, by tacking on to the qualifications for a Ph.D. an examination in theoretical or practical metascience. Just as the modern philosophy of science tells us that science has no unique cognitive method, so the sociology of science tells us that the norms of the scientific community must not be regarded as mandatory. The very fact that we can perceive a very significant social component in the research process should warn us not to interfere with a heavy hand in the learning of that very delicate craft.

Nevertheless, the transition from undergraduate learning to postgraduate research is a difficult personal experience for many people. One would hope that this shock would be lessened, and the transition made more smoothly, if students already had a much better idea of what to expect. If there were general STS education at school, and in the earlier stages of the undergraduate curriculum, this would certainly help to prepare them for the research professions – even to the point of discouraging a few aspirants with quite unreal conceptions of the true nature of modern science. Beyond this, however, there is still a place in the final year of an Honours degree for a somewhat more specialist course concerning research as a vocation, both for students considering whether to go directly into the R & D system and also for the many others who are likely to be concerned with the products of this system in industry, commerce and government.

A curriculum for such a course is easy to sketch out. It would constitute, so to speak, a 'personal' approach to STS studies concentrating on the relationship of the individual scientist with professional colleagues, with the communication system of science, with research apparatus, with administrative bureaucracies, with government agencies, with corporate employers, and with the general public. In other words, it would deal with most of the important aspects of the R & D system – psychological, cognitive, sociological, economic, political and cultural – from the point of view of its active participants. If such a course were built upon a good foundation of STS education in earlier years, it could be made quite thorough and scholarly, especially in the discussion of those deep questions concerning the grounds for belief in the results of research which are at the very heart of science. In that case, perhaps it should be optional. But there could scarcely be a more rewarding intellectual challenge to a serious STS scholar than to present such a course to a group of able science students already well on the way to the frontiers of knowledge.

8.7 General science for society in general

Many science graduates take up employment in industry and commerce, or in national and local government, where they make practically no direct use of advanced scientific knowledge. Even those science graduates who become school teachers seldom have occasion in their work to refer specifically to their scientific education beyond A level.

For reasons discussed in chapter 1, the system of science education does not take much account of the real educational needs of this quite large group. The tradition has been to drag them as far as possible up the pyramid of valid science (§2.1), just as if they were preparing themselves for research. But it has always been recognized that their needs are better met by a more general course, in which, for example, a student could take several different scientific (or non-scientific) subjects to a lower level, rather than trying to follow a single discipline as far as he or she could manage to go. The expansion of higher education in the last twenty years has tended to push such courses out of the university sector, but many students in polytechnics and colleges of education are following 'general science' or 'combined science' courses either to Honours standard, or (in the peculiar jargon of modern British academia) to the somewhat lower standard of an 'ordinary' degree.

But mere eclecticism is not enough. There is little educational value in a kaleidoscopic collection of unrelated electives, in which a student may earn a degree by choosing modules for, say, cellular biology, combinatorial topology, cultural anthropology, Coptic, and creative writing. Nor is it desirable to pretend to marry two disciplines that have nothing in common – a 'Joint Honours School in Physics and History', for example, or in 'Literature and Meteorology'. The intention to 'combine' or 'integrate' the studies is fundamental to their educational purpose.

The weakness of the traditional General Science degree was always the lack of any connection between the various subjects, even when these were intellectually cognate (§7.4). Just as at A level, one might attend classes in, say, physics, chemistry, and mathematics, with no mention at all of mathematical physical chemistry, or of chemical engineering, where these separate skills would have had to be combined. Even a student being trained to teach science in the lower levels of a secondary school had to learn each branch of science as a separate discipline.

The spread of subjects in a general degree has great potential value as a medium of education – but only when they have been connected to give a unified view of some broad aspect of nature or of human affairs. In most courses of this kind, it is now regarded as essential that there should be a component of general studies to 'enrich' and 'integrate' the separate subjects. This is the one place in the system of science education where STS education has become firmly established and institutionalized.

It is scarcely necessary to point out how suitable STS topics are for this integrating role. Their whole spirit is transdisciplinary. A problematic approach demonstrates how distinct subjects such as chemistry, genetics and biochemistry must be combined to deal with real issues such as food resources or environmental pollution. The political and economic issues that arise in science policy studies do not fall neatly into the standard departmental pigeon holes, but stretch right across the academic spectrum. Metascientific principles exemplify themselves in the history of all the sciences in equal measure. As in the Sixth Form, STS is ideal as a theme for a course of general studies, bringing together scientists and non-scientists on an equal footing.

From the point of view of the STS movement this is a very attractive field for educational experiment and innovation. The vocational and disciplinary constraints are undemanding: it cannot be insisted that the students *must* be familiar with any particular item of scientific or cultural information, nor that they should have been tuned to a high intellectual pitch, or demonstrated excellence in some special skill. The wide range of possibilities is reflected in the diversity of STS curricula that are in fact being taught or have been proposed for this purpose.

Without a detailed survey, it is impossible to do justice to this sector of STS education. But two major principles ought to be kept in mind.

The first is that the STS course should indeed enrich the whole student curriculum by connecting with and drawing together the separate subjects, scientific or otherwise. It must not seem to the students yet another 'subject', in the characteristic academic sense, divorced from anything they are learning in other departments of the college. This puts firmly on the teaching staff the responsibility for proper interdepartmental cooperation and coordination, to give the whole degree a genuine interdisciplinary flavour. It also governs the choice of topics and pedagogic techniques in the STS course; it is inappropriate, for example, for these all to be drawn from the theory and practice of the social sciences, without any apparent relation to the techniques of the natural sciences that the students have been learning for a number of years.

The second point is that the freedom of the STS course from disciplinary and vocational constraints should not be interpreted as licence for a group of teachers to ride their personal hobby horses (§7.9). It may be that a General Science degree would be enriched culturally by a good solid course about the politics and economics of social development, or about the limits to growth (§7.8), or about industrial management (§5.6), or about the latest neo-Marxist version of the sociology of knowledge (§6.5). Each of these is relevant to the social role of science, and could be the starting point for a valuable educational experience; but none of them can be regarded as so central to the STS theme as to monopolize the course as a whole. If it is supposed to be about science, and technology, and society, then it must treat these topics evenhandedly, and must itself be integrated around the area they have in common. Without a unifying principle such as the model of the R & D system discussed in chapters 4–6, this sort of course is liable to exemplify the fragmentation of viewpoints and the diversity of goals from which the whole STS movement tends to suffer (§4.1).

Needless to say, the STS component of a General Science degree must be able to compete in sound scholarship and intrinsic interest with the teaching provided by specialist subject departments. But there is not the same pressure as in a special Honours degree to stuff them full of technical information. There is quite a lot of freedom within any particular institution to decide the weight to be given to the integrating component within the degree as a whole, so that no standard pattern has yet evolved.

Some institutions, for example, go little beyond trying to insert items of 'relevance' within the separate subject courses (cf. §7.2). In other colleges, the STS course is merely one option amongst a number of parallel courses provided by a department of General Studies. It is rare for it to be allowed to take up a significant fraction of the whole degree curriculum, and thus to be taught up to a high level of academic sophistication. Nor is there always a response to the vocational element in some types of general degree. Even in colleges of education, where the STS component in a B.Ed. degree could be directed towards the growing involvement of science teachers in STS education in the schools (§8.3), this opportunity is seldom satisfactorily exploited. Again, in a course with a strong admixture of industrial and commercial studies, there may be little attention to the innovative, instrumental, and economic aspects of the R & D system, which would obviously be relevant for careers in industrial management.

This is a sector of higher education where the aims and objectives, the syllabuses and curricula, the teaching techniques and methods of student assessment, are all in a state of flux. The particular problems of STS education will no doubt sort themselves out in time. Well-conceived courses will survive, and those that are ill-matched to student needs or interests will die. It is vital for the STS movement both to guide these developments thoughtfully – and to learn from the experience of those who have been most deeply involved.

8.8 First degrees in Science Studies

It seems a short step from the General Science degree to a course in which STS studies are the major component. But the place of more or less specialized 'Science Studies' degrees within academia, and within STS education, is more complex than appears at first sight.

In one important respect, experience with such courses has been discouraging: student enrolment has been much weaker than was hoped originally. On the face of it, this is surprising. To a mature academic observer, or to the recruiting teams for large-scale industry and government, a graduate with a good Honours degree in this sort of subject would seem to be very well-qualified for a wide range of administrative posts. There might not be the overwhelming evidence of single-minded intellectual power that is supposedly measured by Honours examinations in one tough discipline, but a course in Science Studies would exercise and make evident many other valuable gifts of mind and character. In fact, the argument for such courses as a liberal education for 'generalists' exactly parallels (§7.4) the well-entrenched case for Oxford 'Greats' and 'Modern Greats', which are really no more than general degrees integrated around classical humanism and contemporary socio-political thought, respectively. It could easily happen, in the next ten or twenty years, that the Science Studies course at a good university becomes fashionable as a 'sound education' in an 'intellectually demanding subject', making it a qualification for an up-stage job, and hence a highly competitive educational crucible for very able students. From this successful exemplar would follow popularity for many similar courses throughout the higher educational system.

Nevertheless, neither school children nor their teachers and career advisers yet perceive these attractions. Most likely, this is because Science Studies have not yet acquired an academic identity. It is a novel subject, not clearly associated with a familiar discipline like history or geology. It is not like biochemistry, say, simply the interdisciplinary area between two traditional disciplines. It does not apparently lead into a well-defined professional niche, like social work or accountancy. It is not even the happy hunting ground for a few enthusiasts, like the history of science, or social anthropology. Until STS education becomes fully established as a distinct subject at school level, students will be very suspicious of it in higher education because they have no idea of what it is about nor where it might lead them.

This raises the very general question, to be discussed further in chapter 9, whether STS should now be regarded as the subject matter of an academic discipline in its own right. So long as STS is studied in subsidiary courses by students whose prime allegiance is to other subjects, this question need not be faced directly. Up to the level that we have discussed till now, it is positively misleading to impose the special terminology and deep theory of any of the relevant disciplines upon what is being taught. It is the duty of the STS teacher to accommodate the exposition of each particular view of the subject to the general knowledge and understanding of his or her students. There is much value in *demonstrating* to the narrow-minded science student the capabilities of a sociological or humanistic attitude, but these courses are never intended as instruments for *inculcating* quite unfamiliar ways of thought. In the same way, as a medium of general education for the non-scientist, STS education is *about* science, and does not instill scientific knowledge as such.

But this vagueness cannot be justified as we increase the amount of time that a student is expected to put into this subject. As we look more closely into various aspects of the R & D system, we begin to ask questions that cannot be answered by an appeal to folk knowledge or elementary common sense. Questions such as 'What are the certain foundations for belief in science?' or 'Who should prescribe the objectives of science policy?' or 'Why is long-term research not necessarily a profitable investment?' lead into the central areas of philosophy, sociology, politics and economics. At a certain point we are bound to take account of high scholarship in these various disciplines, and before we know where we are we find ourselves talking the language of a genuine departmental subspecialty such as 'philosophy of science', or 'sociology of science' or 'the economics of research and innovation'. Whether or not there ought to be a distinct STS discipline, there is no doubt that Science Studies must ultimately be intellectually disciplined.

This puts severe strains upon all those concerned with Science Studies as a major degree subject. For the students there is the formidable task of trying to assimilate the key concepts and terminology of several antipathetic academic tribes. For each teacher there is a tension between scholarly loyalty to a particular departmental upbringing and the educational responsibility to adopt a broader view. For the institution establishing such a course there is the problem of collecting a sufficiently diverse team of teachers to cover the various branches of the subject and giving them a home together in the faculty structure.

In the extreme, an undergraduate degree course concentrated solely on 'Science, Technology, and Society' would certainly ask too much of students and teachers alike. There are great difficulties even in setting up balanced 'Science Studies' degrees where these topics take up only a fraction – say a third or a half – of the whole curriculum. Apart from STS general studies, what we mainly find at first degree level are specialized courses on various aspects of Science Studies as major options in existing degree syllabuses, in various relevant disciplines. Thus, for example, some departments of philosophy now specialize heavily in the philosophy of science; some departments of economics and of management teach about technological research and innovation; and so on.

These courses normally arise out of the specialized scholarly and research interests of particular groups of teachers, and are usually much more sophisticated than anything that can be taught in an STS course in a general science degree. It is obviously very attractive for an enthusiastic academic research worker to be able to teach his subject at a high intellectual level, applying the full weight of a particular disciplinary mode of thought. But they are not a satisfactory model for wider degrees in Science Studies. The plain fact is that the educational achievements of such courses are not encouraging. As things stand at present undergraduate students are simply not prepared for the full rigours of the contemporary metascientific disciplines. They have neither the knowledge and experience of science itself nor of any single discipline in the social sciences to appreciate the limitations of what they are being asked to learn. They are expected to acquire a superior understanding, in the abstract, of subjects on which they are profoundly ignorant in concrete terms. What they come out with is not merely immature or inaccurate – it is often grotesquely distorted or absurd.

What can easily happen is that students of philosophy, say, or of sociology, with no more than A-level knowledge of the contents of any branch of the natural sciences, with absolutely no contact with basic research or technological practice, are taught about science entirely from a philosophical or sociological point of view. They are zealously encouraged to examine and comment on, say, the detailed forms of logical justification of advanced scientific theory, or the extent to which communal norms govern the behaviour of research scientists, with only fragments of empirical knowledge on which to base their opinions. Since these issues are controversial at the highest scholarly level, and are usually expounded with honest partisan conviction, it is scarcely surprising that students come out of such courses with bizarre misconceptions, such as that the sole object of research is to falsify theories, or that all scientific theories are arbitrary and relative, or that all research is directed towards the support of the class structure, or that the scientific community maintains itself simply by the exchange of 'contributions' for 'recognition'.

It is not fair to blame the students: they have no defences of fact or faith against such half-baked doctrines. And perhaps one should not be too hard on teachers who are sincerely attempting to express their own sophisticated vision of what they know. But in their zeal to correct the amateur viewpoint of much that is written about science and society, they fall into the very sins of scholasticism, dogmatism and intellectual élitism of which they accuse the esoteric disciplines of 'valid' science.

This is a warning against rushing students into the higher fields of academic Science Studies without sound preparation at more elementary levels. The first priority of STS education is to establish a proper curriculum that gives a good general idea of the shape of the subject as a whole, and to get this accepted in schools, polytechnics and universities, not only as a means of enriching science teaching but also as an essential preliminary to more advanced work in science studies. Until this has been achieved, the attempt to teach about science in metascientific abstractions (§§7.6, 7.7) can bring the STS movement into disrepute amongst liberal minded scientists and technologists, and is probably doing more educational harm than good to those who are subjected to it.

8.9 Postgraduate STS teaching

The place for disciplinary-oriented STS teaching is at postgraduate level. It is usual in academia for candidates for research degrees, such as the Ph.D., to undergo instruction on advanced topics related to their research problems. But the real need here is for systematic courses with broader objectives than is customary for research students in most branches of 'valid' science. In other words, there is a very important place in STS education for courses of study at the Master's degree level, integrated with, but not subsidiary to, original research.

There are already several admirable examples of such courses. Those who undertake such courses do so not so much to qualify for advanced research in STS subjects, or as a way into the upper reaches of academia itself, but in general preparation for a variety of possible careers in higher technological management, in science journalism, in school and college teaching, in the civil service, in technical library and information services, in scholarly publishing, etc. Even here, at their most rigorous, STS studies retain their value as a medium of general education.

Postgraduate courses in science studies not only give the lead and set intellectual standards for STS education at all levels; they are also important gateways between the sciences and society. The separation of the 'Two Cultures' is universally deplored; advanced study and research on the 'humanities of science', on the 'politics of technology', on the 'sociology of knowledge' bridges the traditional faculties and introduces into society at large people who are cultured from both points of view. If we are to continue to be governed by a meritocratic élite, then this élite needs STS education even more than the majority of citizens.

It is surprising, at first sight, that sophisticated courses in science studies work so much better at the postgraduate level than they do even in the final year of a first degree. The reason is that they are being taken by groups of students who are much more diverse in their backgrounds and much more interested in the subject as such. They have seldom been specially prepared in STS themes: graduates in the various conventional sciences are admitted along with social science and education graduates, usually including a few, more mature persons with industrial, technical, teaching or administrative experience. They may not know much about science in its social context, but they have the curiosity and the confidence to question their teachers, their fellow students – and their own preconceptions. They may be assumed to be educationally self-winding, fully capable of organizing their own reading and writing and of bearing the responsibility for their own successes and failures. They are on much more equal terms with their teachers than undergraduates: they are less of an educational burden, and more of an intellectual challenge.

Under these circumstances, there can be no question of laying down guidelines for the contents or conduct of such courses. There can be no pretence that a Master's degree, Diploma, or Doctorate in a branch of science studies certifies any definable expertise or technical skill. It is not like a qualification in engineering, or medicine, or in a very specific field of applied science, where the actual subject matter bears directly upon professional employment. The topics to be studied, the techniques of instruction, the forms of examination, the relative weights to be assigned to class work, essays and research projects must be decided by the teaching group. In the long run, the quality of such a course can only be measured in very subjective terms, against intuitive standards of educational and intellectual value.

There is still the danger, nevertheless, of talking sagely about some aspect or dimension of science, technology, and society without reference to empirical realities. The virtues of an eclectic multi-disciplinary approach must still be recognized. Practically speaking, it is fairly easy to collect together enough teaching power to mount a one year Master's degree course for a small group of students. When that power can be found already concentrated in a single disciplinary department, there is a strong temptation not to look outside for other teachers, other points of view, or students with other backgrounds. The arguments for founding STS education upon a broad model of the R & D system are as cogent at the postgraduate level as they are further down the academic pyramid. Indeed, it is at this most advanced and intellectually sophisticated level that this model needs to be made most explicit, as a framework within which each controversial topic can be shown to have a place, and as an integrating theme for the whole subject.

9

Getting STS education established

9.1 The cycle of innovation

The educational system in Britain – perhaps it is the same everywhere – has a characteristic penetration time of about twenty years. This seems to be about the time it takes for an educational innovation to get properly established. When it is first mooted, by one or two unconventional thinkers, it is dismissed out of hand by the official authorities, who point out how unnecessary, damaging or absurd it would be to, say, raise the school-leaving age, make all secondary education comprehensive, teach children to use electronic calculators, or whatever else new is suggested. It then takes at least a decade of seemingly fruitless debate, shoestring trials, heart-breaking set-backs, inconclusive investigations and other disappointments before the enthusiasts prove their point in principle.

This, in my opinion, is the present state of the movement for STS education. It has been going now for something like ten years, and the pundits of science education are at last beginning to concede that there is a good case for some sort of reform of this kind. Indeed, as is the way with pundits, they can most of them now argue the case for it as if they had never thought otherwise, quite ignoring their previous scornful opposition.

But this is only half way through the cycle of innovation. The debate must now shift from questions of principle to questions of practice – where are the resources, the teachers, the textbooks, to do what we all agree (don't we?) ought to be done? What should be the curriculum, who will set standards, will there be examination syllabuses, etc., etc.; the authorities fight a rear-guard action in a profusion of questions calculated to daunt the 'well-intentioned' but 'hopelessly unrealistic' innovator.

As it enters this second phase, the STS movement must expect to face quite a new range of problems, for which it is very ill-prepared. It is one thing to argue amongst sympathetic colleagues about whether one should make a distinction in principle between basic science and technology, or to press the case in a Faculty Board of Studies for practical projects in place of written examination papers; it is quite a different matter to make decisions affecting the abiding personnel problems of academia – career ladders, promotion prospects, research opportunities, departmental affiliations, etc. – in the Committee of Deans or in response to a CNAA visitation. The task of actually getting STS education *established* within the educational system is not going to be nearly as much fun as showing that this development is both desirable and feasible. And judging by the history of other educational innovations, there at least is another decade of very hard work and careful thought before this goal will be within reach.

The issues that are likely to arise are not only tightly knotted and harshly material; they also penetrate so far into the tissues of the whole educational enterprise, and touch upon so many nerve points of pride and prejudice, sense and sensibility, that they are not to be resolved by sage philosophizing or energetic slogan-mongering. Pragmatism, patient diplomacy, imaginative compromise and suchlike creative arts of social engineering will be the only practical means of further progress.

But one can discern a few general issues around which numerous specific points of debate are likely to turn. A great deal of the success of STS education must depend upon the skills of those who teach it – that is, upon the education that they have received in training for this particular vocation. To discuss this question seriously, one must consider the relationship between the different forms of STS education, at different levels in schools and colleges, and the extent to which these should be regarded as belonging to a single academic subject – perhaps to be differentiated as a separate 'discipline' and organized into a separate 'department' within each educational institution. These are the delicate issues to be taken up briefly in this final chapter.

9.2 Training for STS teaching

Up till now, STS education has had to rely upon a very diverse collection of teachers. There is a professional nucleus of people with postgraduate degrees in one or another branch of science studies. There are other youngish graduates in the conventional sciences or other disciplines who have trained themselves to an adequate scholarly level in STS topics. There are a few experienced school-teachers and academics who have moved seriously into STS education from other fields such as sociology, philosophy or economics. There are quite a number of conventional science teachers who have become enthusiastic about some aspect of STS education, and who give themselves zealously – if rather amateurishly – to it as a good cause.

For the time being, these are the people that drive the STS movement, and on whom it must depend. It is more important to bring them together in a unified endeavour than to discriminate between them on the grounds of academic qualifications or educational experience. Until a more or less standard STS curriculum has evolved out of the present diversity of experimental schemes, it would be wrong to insist that it should only be taught by licensed experts.

Many STS enthusiasts positively rejoice in this anarchy. Nevertheless, we should be looking forward to a time when STS studies have become established as a normal component of the educational system at secondary and tertiary levels. It will then have to be taught with the same professionalism as any other subject. There will have to be a recognized scheme by which STS teachers are trained, and a career structure within which their personal aspirations may be satisfied. Every educational innovation is as much constrained by the personal motives and perceptions of its regular teachers as by the capabilities of its ordinary pupils. In the long run, the form and social influence of STS education will depend on the institutional patterns in which it becomes stabilized, and especially on the wheel of instruction, from teacher, to pupil, to the next generation of teachers and pupils, by which it is sustained and carried forward perennially.

Suppose, for the sake of argument, we envisage STS education on quite a large scale, throughout the whole educational system, in schools, universities and polytechnics (§9.5). In the previous chapter, we have tried to work out appropriate approaches and curricular objectives for each of the main streams of scientific and technological education, at each level; how would the necessary teachers be trained and employed within our existing institutional framework?

The basic principle of professional training for all branches of education (§1.6) is that the teacher trainee should complete at least one cycle of instruction at a higher level than the one which he or she is to teach. Thus, for example, a specialist science degree is the essential qualification for teaching the corresponding subject up to A level, whilst a university or polytechnic lecturer teaching to Honours level should have the equivalent of a Ph.D. in postgraduate study and advanced research. Whilst resisting pressures to rigidify this principle into a strict regulation, one must accept it as both a wise maxim and a practical convention of academic life.

This principle has a very significant consequence – that education in any particular subject is stitched together, from level to level, by the return of graduates from higher levels to teach in the more elementary cycles. We have already noted the effects of this in 'valid' science subjects, where both the high critical standards and the puristic scientism of advanced basic research are transmitted right down the pyramid into the earlier years of the secondary school (§2.6). We may anticipate similar effects in STS education when it has grown to maturity as an academic subject.

What this means, for example, is that the approach to relevance through interdisciplinary teaching of science in the lower and middle school (§8.2) depends upon the proper training of science teachers along those lines in colleges of education, polytechnics and universities (§8.4). Similarly, the quality of STS general studies in the Sixth Form (§8.3) will be determined by the style and contents of the STS courses taken by specialist science teachers – for example, by the STS components of science degrees and/or postgraduate certificates in education (§8.7). Undergraduate courses, in their turn, will have to be taught by people with more advanced qualifications, such as Master's degrees in science studies (§8.9).

Training for STS teaching cannot be separated from conventional science education. It is impossible to teach *about* science without a knowledge *of* science up to some minimum level of validity. An STS teacher who clearly knows less of the basic elements of a science than his or her pupils is not going to make much impression on them. As a general principle, it might be held that no STS course should be taught by some one who has not passed through a course of 'valid' science at the same level – for example, to teach about the social context of physics to physics undergraduates one should at least have taken a first degree in one or several physical sciences, or have had equivalent experience of scientific work. The supposition that advanced study or research in a metascientific discipline can substitute for some formal education in the language, concepts, methods and approach of a particular 'valid' science is not borne out by any direct observation of academic life.

Indeed, most scientists, and at least a few historians, sociologists and philosophers of science, would regard a solid scientific education as an absolute prerequisite for teaching and research on STS themes. It is often argued, further, that serious scholarly study of the R & D system can only be done properly by people with personal experience of research, at least up to the Ph.D. stage, in 'valid' science or technology: how can one really understand what it means to make a scientific discovery, refute a theory, referee a paper, take part in a scientific controversy, etc., without having been directly involved in these mysterious transactions of the scientific life?

This argument is, of course, a special case of the standard debate on the relative merits of 'insiders' and 'outsiders' as social observers. Anthropologists, for example, pose the question: 'If you are not an Ojibway by upbringing, how can you possibly know what it is like to be one?' against the equally valid proposition that only someone who is not an Ojibway can be sufficiently detached to see what Ojibway society is really about. This dilemma applies to all social and humanistic studies, and cannot be avoided by going to one or the other extreme. Neither the insider nor the outsider has a monopoly of insight or knowledge. It is an empirical fact that very eminent scientists can philosophize and sociologize about science as ignorantly and naively as quite lowly philosophers and sociologists without direct experience of research. The actualities of a scientific career cannot be rejected as evidence in the name of abstract metascientific theory, but it is the factual quality of this evidence that counts, not the status of the person who pontificates about it. The claim that STS teaching must be kept exclusively in the hands of mature research workers and science teachers cannot be accepted; the contrary tendency to separate education in science studies from conventional science education must also be firmly resisted.

In the end, the healthy development of STS education at each level depends upon the advancement of the whole subject, from the grass roots of teaching the general citizen about the place of science in modern life to the academic tree tops of metascientific learning and research. Channels of influence flow both ways through this system, up from the vocational needs of pupils and students and the foundations that are laid for their later studies, and down from the more informed and up-to-date material of advanced scholarship, with its direct involvement in contemporary issues of principle and practice. Nothing could be more damaging for the STS movement than to cut it into slices and segments, separating school teachers from academics, polytechnics from universities, general studies from specialized degrees and postgraduate courses, or science policy from history, philosophy or sociology. The underlying unity of the subject in its educational aims and intellectual themes must be reflected in a coherent framework of curriculum development, teacher training, and professional employment within the educational system as a whole.

An educational curriculum is a social institution; to keep alive, it must enjoy the loyalty and personal commitment of energetic and enthusiastic supporters, by whose efforts it is continually being renewed. The STS movement draws this vitality from both the bottom and the top of the educational pyramid – from the changing needs and aspirations of those to be taught and the scholarly drives of its most learned teachers. The call for intellectual rigour and 'validity' in science studies is not based solely upon a desire to clothe this new subject in superficial academic respectability: it derives equally from the truth that it is from the nuclear regions of a discipline that genuine progress normally flows. The whole of STS education would degenerate into amateur dogmatics if it were not continually illuminated by the results of serious scholarship and research.

The whole question of the objectives and achievements of advanced study and research on the social relations of science is much too large to be dealt with in this book. It is enough to say that there is no end to the potentialities of STS as a field of investigation for sociologists, social psychologists, political scientists, economists, philosophers, social historians and many other contemporary academic breeds. This research justifies itself by its contributions to science policy making, and by the more reliable image it can give of the role of science and technology in all spheres of material culture, ideology, education, and social action. The constraints are shortages of properly trained people, and of financial support – not of problems worthy of investigation.

The relevant point here is that STS research must be closely linked with STS education at the highest levels. As in every serious academic subject, advanced scholarship goes hand in hand with advanced teaching, for their mutual benefit. It is just as important that research workers should be made to teach what they think they know as it is for teachers to stretch their minds by participation in research. The characteristic research topics of the advanced metascientific disciplines are usually somewhat too sophisticated for first degree courses. As was emphasized in §8.8, the average undergraduate student in science, technology, or social science is too ignorant about the R & D system, and the social role of science in general, to make good sense of such topics. As in any conventional academic discipline, it is a great mistake to stuff the minds of undergraduates with scarcely comprehensible research material in a manner that would be only appropriate in a postgraduate course. But this is no excuse for an STS scholar to be unconcerned about STS education. By its own theoretical principles there can be no division of labour between researchers and teachers in this field of studies. The STS scholar who, say, preaches social responsibility in science, or who demonstrates the importance of the scientific communication system, or who emphasizes the function of the norms of communality and universality in the scientific community, or who stresses the need to challenge theoretical paradigms, must surely appreciate that these responsibilities and norms apply to STS itself, in its educational activities as much as in its highbrow intellectual forms. Not all teachers are in a position to undertake original research; nevertheless, there is an onus upon every research scholar to communicate the essence of recent discoveries and insights, to the very best of his or her ability, directly to students themselves and indirectly to those who teach the subject at all lower levels.

9.3 STS teaching in the schools

In the previous chapter, it was suggested that the approach to STS education in the lower and middle school should be primarily through greater relevance and interdisciplinarity in the teaching of ordinary science subjects (§8.2). If that is accepted, then the teachers of STS topics can be none other than the ordinary science teachers. Indeed, the whole thrust of the STS movement at this level is not towards the introduction of a completely new school subject but rather to modify and reorient science education itself in a variety of ways.

But to make progress in that direction it will not be sufficient to devise a few new curricula and examination syllabuses. What we are really seeking is to establish a new tone of voice in all science teaching (and examining) at the pre-certificate level. This attitude will need to diffuse throughout the whole profession. STS education in the lower and middle school is not a 'subject', like Home Economics or Greek, to be entrusted to a few specialists: it must be part of the repertoire of knowledge and skill of every science teacher, to have the wit and will to seize upon opportunities that may arise in any lesson on any biological, mathematical, environmental or physical topic.

How is this new orientation to be achieved? There will obviously have to be quite a lot of hard thinking, serious discussion, self-education and re-reducation amongst science teachers. But the amount of really sophisticated special knowledge required is negligible. It is absurd to think of training them as STS specialists, as if they were expected to give lessons on the role of the Academie des Sciences in 19th century France, or on Lakatos's concept of a research programme, or on the computational deficiencies of *Limits to Growth*. It is not much for a physics teacher to learn some of the engineering aspects of the energy principle, or for a chemistry teacher to explain how plastics were discovered, or for a biology teacher to escape from the innards of the earth worm to show how diseases are transmitted by insects in tropical countries. This attitude towards the more relevant applications of the conventional disciplines is nothing very new or difficult to foster in science education as traditionally practised.

In fact, the greatest obstacle to this whole trend in science education is to establish adequate interdisciplinary competence in the conventional sense. Teachers with physics degrees may have never learnt any biology; biologists are often totally ignorant of the most elementary physics. The price of over-specialized higher education in the 'valid' science disciplines must now be paid. If fully integrated, transdisciplinary courses are to be taught up to CSE/GCE O level, there will have to be a revolution in the basic training of science teachers in colleges of education, polytechnics, and universities. Until this particular revolution is well under way, the STS emphasis may have to be on relevance, rather than on interdisciplinarity as such.

The movement towards more socially relevant science teaching up to this level is hindered less by real ignorance on the part of school teachers than by sheer educational tradition. This prejudice is already melting away. STS topics are now beginning to be taught in colleges, departments, schools and institutes of education, as part of the professional training of science teachers. Indeed, this vocational need is often used to justify the inclusion of a corresponding amount of STS material in conventional science degree courses. 'The history and philosophy of science is useless for real research', say the science professors, 'but it's just the sort of thing that school-teachers should know about!' Despite doubts concerning the quality of much that is at present taught under this heading, we may have some confidence that future science teachers can be given sufficient contact with ideas about the social context of science not to be frightened of talking about such things to their pupils.

Nevertheless, setting the tone of elementary science education for the general public could be a real challenge for the STS movement. It cannot be done with ponderous academic metascience, nor by fervent social concern. If new curricula are to be devised, they simply must be workmanlike and teachable to the vast mass of ordinary children in ordinary schools. If books, worksheets, teachers' guides and similar auxiliary material is to be produced, it simply must be properly tested in the classroom according to the same criteria. The great mass of serving teachers will need a great deal of help, not only with such conventional educational paraphernalia, but also in rethinking their own practices, the real needs of their pupils, and the conflicting demands of the society in which they live.

The same goes for the trend towards greater social relevance in the teaching of science subjects in the Sixth Form, whether in preparation for entry to college, as a qualification for technical employment, or simply as a part of general education. But, as we saw in §8.3, this trend cannot be carried too far without compromising the 'validity' that is essential in science education. The full range of STS topics can only be brought satisfactorily into the Sixth Form curriculum in the form of General Studies that are complementary to the standard Science and Arts A-level subjects.

Who, then, should be teaching these courses? It would be inconsistent with their marginal, complementary, general-educational purpose if they could only be taught by highly qualified specialists with advanced degrees in science studies. The whole idea is surely that these are not strictly academic topics, subject to the usual criteria of rigour and validity, but are open-ended and culturally enriching, relying more upon the wisdom, experience and intellectual maturity of the teacher than upon formal book learning. In other words STS teaching in the Sixth Form may call for a particular type of person, a good range of relevant general knowledge, and an appropriate style of teaching, rather than any elaborate special training. One might expect to find these qualities in any large school amongst the more experienced and senior teachers of the conventional Science subjects, and in at least a few of their colleagues in the Humanities and Social Sciences.

It must be admitted, however, that very few of these potential contributors to STS education are yet sufficiently prepared for this particular task. In the long run, we may anticipate that the various more advanced forms of STS study at college level, especially where these are complementary to 'valid' science in science and education degrees, will provide this background for future school teachers. We may also expect, as the subject establishes itself in the schools, that a number of science teachers will seek to qualify themselves more professionally by taking courses at Master's level – for example, for the M.Ed. degree – with a significant STS slant.

In the meantime, however, until a corps of competent teachers has been created, there will be a great need for a variety of initiatives – curriculum development projects, production of textbooks and other teaching materials, short courses and in-service seminars, etc. – to encourage science teachers and others to take up this subject in the upper levels of the secondary school. It is not so much that there is a great deal to learn, or that difficult new concepts must be mastered, but that guidance is needed through the diffuse and ill-assorted literature, to grasp the main features of science in its social context and to see how these can be related to the interests and capacities of children at school. This is where the more expert STS scholars in the higher reaches of academia have a vital role: by stepping outside their narrow university and polytechnic circles, they can give a valuable intellectual lead in STS education at school level as it struggles to define itself and clarify its objectives and methods. As I have emphasized throughout this book, in this field it is not a matter of setting up standards of 'validity', but of making quite clear what the subject is really about, the various ways it can be approached, and how its various aspects can be articulated and integrated. Many science teachers are persuaded of the importance of STS topics, and are enthusiastic to introduce them into their schools; but they are not very confident of their own competence to teach in unfamiliar ways, and are waiting for this leadership before they can begin.

9.4 STS teaching in tertiary education

At college level, in universities, polytechnics and colleges of education, STS studies may take many forms – as complementary or integrating studies in General Science degrees (§8.7), as an enrichment of a special Honours degree in science (§8.4), as a preparation for the technological and research professions (§§8.5, 8.6), or as a major component of a degree in philosophy, sociology, economics or politics (§8.7). There is no doubt that suitable curricula could be prescribed for these studies, including appropriate examination syllabuses, or other means of assessing student progress, if so desired. But who would give the lectures, moderate the seminars, guide the case studies and research projects, criticize the essays and mark the examination papers?

It is important to understand clearly that STS teaching at undergraduate level is a distinct educational activity. The situation here is not quite the same as in the Sixth Form, where the STS material needs only to be put together in an orderly manner to be within the scope of any well-educated and experienced teacher. At a higher level, it calls for a new breed of academic: a person who is well acquainted with science studies over a fairly wide field, and who also feels sufficiently at home in a particular scientific discipline not to seem a stranger to ordinary science students. It cannot be done satisfactorily as a sideline, either by conventional science teachers or by scholarly experts in the various cognate disciplines, such as philosophy or sociology. For the moment, we may have to be satisfied with self-educated STS experts, who have taken up the subject out of interest, or moral fervour, or to fill a vacancy in the faculty lecture list. In the long run, however, the normal forces of academic life will demand that this subject should be taught by professionally qualified persons, with at least the equivalent of a Master's degree in science studies to certify their competence. Indeed, there is a danger that nothing less than a Ph.D. on an appropriate research topic will ultimately be regarded as a sufficient qualification in the higher status institutions. In other words, the basic training of STS teachers in tertiary education depends upon the existence of specialized science studies courses at the postgraduate level (§8.9), and hence to the establishment of this subject as a distinct specialty within academia – a point to which we shall return in the final sections of this chapter.

In some ways, the question of how college STS teachers should get their initial training is much more straightforward than how they should be placed in the departmental structures of universities and polytechnics. Within the traditions of academia, this sort of question has only two alternative, mutually exclusive answers. On the one hand, STS courses are given within each conventional department – as it might be, of chemistry, or agriculture, or production engineering – by staff 'belonging' to that department; on the other hand, STS studies are made the responsibility of a special department, whose academic staff teach it as a 'service' to students of other departments in one or more faculties. What are the arguments for, and against, these two alternative patterns?

Where the degree structure is strongly departmentalized – that is, where most students are taking special Honours degrees in separate disciplines (§§8.4–8.6) – there seems, at first sight, much in favour of the first pattern. If a significant proportion of the ordinary teaching staff were appropriately trained in STS studies, then this particular duty could be shared out amongst them in the ordinary way: some of the lecturers in physics or chemistry, for example, would be expected to run classes and project work on 'science and society' as part of their job, in the same way as they would lecture on thermodynamics or supervise laboratory work. They would thus be closer in background and interests to their students than if they came in as STS experts from some other department.

This is much the easiest pattern to fit into the conventional structure of academia – but what would be the personal incentives for an academic in a 'valid' scientific or technological discipline to make the considerable effort to become properly qualified in STS studies? There would also be the disadvantages of fragmentations of the whole STS effort in a particular institution amongst individuals who were isolated both from colleagues with similar interests in other departments and from the sources of the knowledge they were trying to expound.

Another possibility is to employ a small number of STS experts as specialist teachers to run such courses, alongside the ordinary academic staff in each department. This would keep the STS teaching within the conventional faculty boundaries, and would not call for a large exercise of retraining existing scientific specialists in these new-fangled subjects. But there is very little attraction for a fully qualified STS teacher in an academic post where there would be practically no opportunities for serious scholarship or research in his or her own field, and no obvious path to promotion as a result of such activities. The real difficulty with the intra-departmental pattern is that STS studies are almost certain to lose momentum and die if the enthusiasm of their teachers fails to earn them recognition in an environment exclusively devoted to the pursuit of a more 'valid' science.

Is it better, then, to set up a separate department of science studies, providing general and specialist STS courses for the whole institution? This has obvious advantages, as the stereotype of the academic division of labour and disciplinary autonomy. Such a department would come to know its scholarly business well, and would provide a congenial home base for its staff. But we also know its deficiencies – the separation of the STS components of the curriculum from what is learnt in other departments, inadequate rapport with students in a multitude of disciplines, the trend towards highbrow topics in what is supposed to be general education, differentiation and specialization of such departments by attraction to the major disciplines of social studies and the humanities, and so on. STS studies would thus tend to develop into yet another academic discipline, available as an optional subsidiary subject for a small proportion of students – a frustrating outcome to the effort to set up such departments in every institution of higher education to enrich scientific and technological education for the mass of students.

In the polytechnic tradition, the STS teaching staff is often concentrated in a department or faculty of general studies, in parallel with other subjects from the humanities and social sciences. This may well be a realistic framework for the actual educational task in many such institutions, with their characteristic patterns of teaching, administration and promotion prospects – for example, in providing complementary or integrating studies in General Science degrees (§8.7). There may be better employment prospects for the STS specialist in this sector than in institutions heavily committed to special Honours degrees in the traditional scientific and technological disciplines, but only at the expense of heavy teaching loads and poor scholarly facilities. In any case, one must have one's doubts about how these patterns will survive under the apparently irresistible pressure to model polytechnics on universities, where the very notion of general education has little influence.

These disheartening consequences of either of the conventional ways of establishing STS studies within the faculty and departmental structure of higher education are not fanciful: they are already observable in particular instances. They arise inevitably from the irreducible tension between general education and special expertise. This contradiction cannot simply be wished away, or deemed to be resolved by an artful compromise. The problems of professional identity and personal preferment to which it gives rise will be further discussed in §9.6.

9.5 The rate and scale of change

Suppose, nevertheless, that the development of STS education is not going to be too severely handicapped by institutional, professional and other structural constraints; to what scale of activity might it be expected to grow, and at what rate? The answers to such questions can be no more than wild conjecture, but it is instructive to make a few simple estimates of the logistics of the operation. In particular, let us suppose that the STS component of higher education in science and technology eventually takes up about five per cent of the time of all students – or, equivalently, that about five per cent of the teaching in faculties of science, engineering and medicine will be done by academic staff with postgraduate STS qualifications. Allowing that these qualifications need not, in general, be so advanced as for specialized instruction in a 'valid' science, we arrive, nevertheless, at a very considerable load on courses in science studies at postgraduate level (§8.9). If, for example, we assume that it takes only one year to put an STS specialist through a Master's degree, instead of three years for the usual Doctoral degree, then at any one time there would have to be about one STS graduate student for every 60 research students in other scientific disciplines. This does not sound a large proportion – but aggregated over all academic departments, and over the whole system of higher education, it amounts to a very substantial activity. At a rough estimate, courses of this kind would need to be established in 20 or 30 different universities and polytechnics to meet such a demand.

Since there are now, at most, about half a dozen viable graduate courses in the general area of 'Science Studies' in our system of higher education, there is not the faintest prospect of establishing STS education on this scale in the next few years. The expansion of an academic discipline is necessarily a very slow process, requiring several decades from the intake of a cohort of students to the maturing of scholars and teachers with the knowledge and experience to direct research and guide advanced instruction. In any case, the material resources are simply not available for such a development.

This logistic exercise, however, is not an idle fancy. It shows quite clearly the great importance of the advanced STS courses that exist at present, as the seed corn for future growth of the subject and as stores of genetic material for any evolution of science studies in other directions. It may also indicate to those who live at these more exalted intellectual levels that *their* work is mainly justified by the contribution it can make, directly and indirectly, to STS education as a whole.

The fundamental lesson of this unlikely scenario is to teach the STS movement to make haste slowly. The human resources that are at present available should not be spread too thinly, each individual STS enthusiast becoming isolated from competent colleagues and liable to frustration and disillusionment in a relatively unsympathetic environment. The way to success may lie through concentration on quality in every aspect of STS teaching – properly thought-out objectives, well-prepared curricula, interesting pedagogic techniques, appropriate books and other material aids – rather than by pressing ahead in all directions supported by the belief that whatever is done must be of value because it is on the side of the angels. Precisely because we have a long way to go, and cannot force the pace, there is time enough to look carefully at what is being done, to consolidate genuine successes (such as, for example, the teaching material produced by the SISCON project), and to cooperate in constructive projects for the future.

9.6 Giving STS an academic identity

In the end, the future of STS education must depend on the skill, the knowledge and the enthusiasm of those who teach it. This book is an attempt to show them that this theme has more coherence of educational intention and academic content than is sometimes realized, and that there is a comprehensible core of knowledge which it is advisable for them to acquire before they set themselves up before a class. Even at the most elementary levels of school science, this competence cannot be taken for granted; at the most advanced levels, STS themes need not necessarily be taught in the pedantic manner of most academic disciplines, but specialized training and scholarly experience will certainly be needed for most of those involved in this task.

The most difficult problem to be faced by the STS movement in the next few years is to develop stable patterns of advanced training, career ladders, departmental affiliations, etc., for STS teachers within our whole system of education. It is the problem of *institutionalizing* this innovation, not only by according it a secure place in the educational curriculum, but also by creating a corresponding social group, an Invisible College, where those who teach this subject can feel confidently at home.

It is natural to look ahead to the standard self-sustaining, self-consistent solution of this problem – that is, where STS is fully established as a major academic subject in its own right. A curriculum incorporating the absolutely necessary concepts of the various relevant disciplines would have grown up. Academics trained primarily in advanced STS studies would feel that they belonged to such a department, and would give their primary allegiance to it, even if their teaching and research used specialized skills drawn from other fields. School-teachers would have taken STS as a major or minor component of their B.Sc., B.Ed., M.Ed., or other degrees, and would thus be formally qualified for employment in appropriately specialized posts. Connections would have been established with practitioners in relevant professions such as research management, technological assessment or science journalism.

This is the hope of many supporters of STS in higher education. But the way things are these days it is not a very realistic prospect. A few small departments of science studies are maintaining themselves primarily by 'service' courses for students of other specialties. History and Philosophy of Science has established itself as an independent academic entity in some institutions, but with little occasion to expand into a full scale discipline offering its own degrees. Neither student push nor employment pull can be cited to justify the large expense of creating entirely new departments devoted to STS. There is not even room, nowadays, to manoeuvre across the frontiers between existing departments to reorient and combine various small groups with complementary interests.

But the recognition of STS studies as a distinct academic discipline is not solely a question of practical opportunities and the availability of resources: in my opinion, the *form* of what is recognized is a vital consideration. As has been quite apparent throughout this book, the subject is of its very essence transdisciplinary. The study of the R & D system involves every department of the social sciences – sociology, economics, politics and social psychology. In the traditional humanities it is deeply involved with philosophy and history. It cannot be followed satisfactorily without scholarly reference to the natural sciences and the scientific technologies. It is not just a subspecialty of an existing discipline, nor does it define itself as an interdisciplinary frontier territory between several different departments. Science is an existential category – something that *is* – which requires to be studied in every possible aspect or dimension, regardless of academic demarcation lines or conceptual schemes.

The core of this subject, whose aspects could not, in principle, be dealt with in one or other of the existing disciplines, is quite small. The scope of highbrow metascience (§§7.5–7.7), for example, is much too narrow to contain all that needs to be thought and taught about science and society. Only a broad definition of the STS field, spreading quite a distance into a variety of other recognized disciplinary areas, can do justice to the subject as a whole. In other words, STS studies cannot be given a distinct academic identity sharply marked off from other subjects. The philosophy of science, the sociology of science, the history of science and technology, the economics and politics of research, etc., etc., must still belong as much to philosophy, sociology, history, economics, etc., as they ever did in the past. Research and teaching in each of these various domains cannot be ruled by a single method or paradigm, but must acknowledge the authority of several different deities at once – including the mighty Pantheon of the natural sciences, from whom all our blessings and afflictions seem to flow.

This sort of intellectual pluralism is not comfortably accommodated in academia. It comes into conflict with the territorial imperatives of departmental and faculty organization. We cannot be perfectly confident that STS studies can be satisfactorily established in this pantheistic transdisciplinary form.

That is why it is so very important for the STS movement to pull itself together voluntarily, outside the official institutional structure of schools and colleges, polytechnics and universities. The scholarly and educational collaboration that is difficult to engineer formally across departmental, faculty, and institutional boundaries will have to be developed informally, by those who are personally involved in these various activities. The identity and integrity of STS studies must come to be recognizable through the activities of learned societies, learned journals, conferences, educational projects, etc., that cut right across the conventional academic categories. It may be only in such activities that philosophers and sociologists, technologists and doom watchers, school-teachers and professors, science historians and science journalists, hard scientists and soft-hearted do-gooders, can really come together for a common purpose.

The future of STS education will probably depend much more on the openness, vitality, and vision of its *Invisible* College than on getting recognition in the traditional manner. Who could better appreciate the social significance and practical possibilities of such a process than those who understand the function and power of just such associations of like-minded persons in science itself.

INDEX